W0037103

Addressing the Climate Crisis

"We shall not get to Net Zero without local action and Local and Regional Authority participation. This book shows how real subsidiarity works and how important it is. It's a clarion call to a Government which still hasn't made local government its partner in the battle against climate change. These are lessons to be learned at every level and their positive story gives us all renewed hope."

—The Rt Hon John Gummer, *Lord Deben, Chairman of the Committee on Climate Change*

"Local communities have emerged as influential drivers of action against climate change and important sources of institutional and behavioural innovation. This powerful book, written by a new generation of climate experts, provides a thoughtful introduction into the theory and practice of place-based climate action."

—Professor Sam Fankhauser, *Professor of Climate Change Economics and Policy, University of Oxford and Research Director, Oxford Martin Initiative on a Net-Zero Recovery*

"What does it mean to deliver place-based climate action, and how can we make that possible? Making a deliberate effort to engage with the complexity of climate action and the contradictions of place-based initiatives, this book points towards the potential for a new, more just, politics of climate change: a politics that puts people's voices at its center."

—Professor Vanesa Castan Broto, *Professor of Climate Urbanism, University of Sheffield and a lead author for the IPCC Working group II on Impacts, Adaptation and Vulnerability*

"There is real urgency to tackle the climate crisis and local action is playing a critical role in this. This book provides a rich set of contributions on the theoretical and practical applications of climate action which make an important contribution to the debate on what it means to deliver and assess action on climate change on the ground."

—Miatta Fahnbulleh, *Chief Executive, New Economics Foundation*

Candice Howarth · Matthew Lane ·
Amanda Slevin
Editors

Addressing the Climate Crisis

Local action in theory and practice

Editors
Candice Howarth
Grantham Research Institute on
Climate Change and the Environment
London School of Economics and
Political Science
London, UK

Matthew Lane
School of Geosciences
University of Edinburgh
Edinburgh, UK

Amanda Slevin
School of History, Anthropology,
Philosophy and Politics
Queen's University Belfast
Belfast, UK

ISBN 978-3-030-79738-6   ISBN 978-3-030-79739-3   (eBook)
https://doi.org/10.1007/978-3-030-79739-3

© The Editor(s) (if applicable) and The Author(s) 2022. This book is an open access publication.
Open Access This book is licensed under the terms of the Creative Commons Attribution 4.0 International License (http://creativecommons.org/licenses/by/4.0/), which permits use, sharing, adaptation, distribution and reproduction in any medium or format, as long as you give appropriate credit to the original author(s) and the source, provide a link to the Creative Commons license and indicate if changes were made.
The images or other third party material in this book are included in the book's Creative Commons license, unless indicated otherwise in a credit line to the material. If material is not included in the book's Creative Commons license and your intended use is not permitted by statutory regulation or exceeds the permitted use, you will need to obtain permission directly from the copyright holder.
The use of general descriptive names, registered names, trademarks, service marks, etc. in this publication does not imply, even in the absence of a specific statement, that such names are exempt from the relevant protective laws and regulations and therefore free for general use.
The publisher, the authors and the editors are safe to assume that the advice and information in this book are believed to be true and accurate at the date of publication. Neither the publisher nor the authors or the editors give a warranty, expressed or implied, with respect to the material contained herein or for any errors or omissions that may have been made. The publisher remains neutral with regard to jurisdictional claims in published maps and institutional affiliations.

Cover illustration: © Melisa Hasan

This Palgrave Macmillan imprint is published by the registered company Springer Nature Switzerland AG
The registered company address is: Gewerbestrasse 11, 6330 Cham, Switzerland

ACKNOWLEDGMENTS

The editors would like to extend their thanks to the Place-Based Climate Action Network (P-CAN) (Ref. ES/S008381/1), funded by the UK Economic and Social Research Council, for supporting this work. Further details of PCAN activities and people can be found at https://pcancities.org.uk/.

The authors of Chapter 3 extend their thanks to Jonny Gordon-Farleigh and David Bollier.

The authors of Chapter 4 thank the participants and members of Envirolution who supported this research with necessary data, knowledge and expertise about their ongoing community education, empowerment and mobilisation project. The award-winning impact assessment report is part of the Collaboration Labs programme at the University of Manchester, funded by the Economic and Social Research Council. Training and coaching support for the research team was supplied by Code Switch Consultants, Future Everything and Inkling Training.

The authors of Chapter 12 would like to acknowledge the diverse stakeholders, project teams and interviewees involved in both projects that frame the reflections for this chapter.

ADDRESSING THE CLIMATE CRISIS: LOCAL CLIMATE ACTION IN THEORY PRACTICE: EDITORIAL

Climate breakdown, extensive biodiversity loss and deepening inequalities—these pressing socio-ecological challenges have profound ramifications for societies, communities and individuals. Inherent to such crises are intersecting socio-economic, ecological and political impacts, which affect people differently and we know that climate breakdown disproportionately affects those with the least resources to adapt, locally and globally (HM Government, 2017). Rather than dwelling on the causes and consequences of unsustainable societal pathways driving socio-ecological adversities, clearly articulated by WMO (2020), IPBES (2019), IPCC (2018), among others, this book offers hope, inspiration and analyses for multi-level climate action, spanning varied communities, places, spaces, agents and disciplines.

At a macro-level, many governments and local authorities have declared a Climate and Ecological Emergency, including the UK (May 2019), Ireland (May 2019), Northern Ireland Assembly (February 2020); by the end of 2020, 79% of District, County, Unitary and Metropolitan Councils in the UK had declared a Climate Emergency (PCAN, 2021). Alongside such declarations, evolving policy frameworks seek to reduce greenhouse gas emissions (GHGs), limit climate change impacts and prepare societies for inevitable consequences of climate breakdown.

Clear policy signals by international organisations and national governments remain essential; however, important decisions are increasingly being made beyond this, and beyond state actors, fostering opportunities for climate action within localities. For example, decisions about low-carbon business opportunities, renewable energy investment, urban transport, energy management, buildings efficiency and the management of climate risks.

It is increasingly recognised that the delivery of climate policy ultimately happens through place-based initiatives at the local level (Galarraga et al., 2011; Howarth et al., 2021), and it has been widely argued that effective delivery of actions to promote low-carbon and climate-resilient development will require experiments with new governance arrangements (Bulkeley et al., 2019; Castán Broto, 2020; Jordan et al., 2018; Kivimaa et al., 2017). In particular, processes that engage and harness the combined energies of public, private and third sectors (Gouldson et al., 2016) are required.

The Place-based Climate Action Network (PCAN), funded by the UK Economic and Social Research Council (ESRC), is focused on translating climate policy into action 'on the ground' in communities. It does this mainly through its three core City Commissions in Belfast, Edinburgh and Leeds (and growing outer network of city commissions), and two thematic platforms looking at sustainable finance and adaptation and resilience, underpinned by business engagement. PCAN collaborated with the Royal Geographical Society's Climate Change Research Group to host a half-day virtual mini conference in September 2020 bringing together international scholars and practitioners working on a range of topics relating to local climate praxis and to discuss the challenges bridging climate theory and practice within place-based climate action.

This book is a result of that mini conference and focuses on applied research exploring the translation of climate policy into action; innovative forms of engagement and citizen participation at the local level and their role in overcoming limited capacity to deliver climate action across different governance levels; diverse forms of 'community' that come to engage with climate action at different scales; critical perspectives on what 'place' means for shaping our response to climate change on different scales; and whether the local context hinders or enhances progress on climate resilience and decarbonisation. We explore these important and pertinent issues to addressing the climate crisis through local climate action with a collection of twelve chapters grouped into three parts: Community and place in local climate action, Spaces of climate action and Agents of local climate action.

Interconnected yet conceptually distinct, the three parts span multiple levels of analysis, interrogating diverse perspectives and practices inherent to the vivid tapestry of climate action emerging locally, nationally and internationally. Setting the scene, Part I illuminates the complexity and value of 'community' and 'place' as a locus for bottom-up approaches to citizen participation in climate mitigation, adaptation and resilience endeavours. Shifting towards meso- and macro-levels of organisations, cities and counties, Part II examines different spaces for local climate action created and/or supported by local authorities. Part III offers insights into key agents of local climate action, demonstrating, on one hand, the centrality of communication and citizen engagement for climate action, and on the other, the important roles played by the private sector and universities.

COMMUNITY AND PLACE IN LOCAL CLIMATE ACTION

When we consider translating climate action into action in communities, we are immediately faced with the basic task of constituting community and place as a starting point for action. 'Community' tends to be seen as the embodiment of activity that occurs in the sometimes-hazy space between the micro-level of individuals and families, and the macro of cities, counties and countries. It can be a catchall term for all sorts of collective activity within civil society yet, conversely, be used to distinguish people as 'insiders' and 'outsiders', excluding those who do not explicitly share bonds at the core of different configurations of community. Similarly, 'place' as a geographical and visceral underpinning for different forms of collective organisation can be levied as subjective or exclusionary. Grappling with conceptual and practical challenges (and opportunities) of community and place, Part I brings crucial elements of local climate action to the fore and offers reflections and shared learning as a basis for impactful climate action that emerges at grassroots level, as opposed to top-down approaches mandated in climate policy.

In Chapter 1, Slevin et al. pose the important question of how do we enable a genuinely inclusive, just transition to low carbon, healthier and fairer communities and societies, particularly in Northern Ireland where legacies of prolonged ethno-nationalist conflict are expressed in physical, cultural and socio-economic divisions. The authors elucidate the potential of community climate action, influenced by Freirean praxis, to tackle injustices, aid peace building and enable a just transition from the bottom-up, when undertaken with and in diverse communities. Concentrating on 'place', in Chapter 2 Murtagh and Lane explore how the concept is understood by place-based climate adaptation projects in Scotland and argue that place is an important, albeit complex and value-laden, mobiliser for collective climate action at a local scale.

Taking a novel approach to collective climate action, in Chapter 3 Stone et al. draw on the work of Elinor Ostrom to situate 'commoning' as an invitation for people to actively participate in the transition to a green economy and society. From their perspective, a Commoners' climate movement involves people claiming, creating and stewarding the infrastructure of a net-zero world through action at multiple scales with multi-level partnerships. In Chapter 4, Walley et al. present innovative empirical data to expound the influence of Freirean critical theory on Envirolution, a Manchester-based volunteer-led cooperative that organises community engagement festivals to raise awareness and inspire action upon the climate emergency.

Combining theory and practice, reflection and action, a consistent thread throughout Part I is the opportunity to engage with understandings of praxis as 'transformation of the world' (Freire, 1996), as a basis for socio-ecological action to aid a just transition to healthier, fairer and more sustainable societies (Slevin et al., 2020), beginning in and with communities and places.

THE SPACES OF CLIMATE ACTION

The 'local' scale has been thrust into the spotlight when it comes to taking action against climate change. Inertia at national and global scales translates into increasing expectation that the local scale can step up and fill the void. But what is the local scale? What spaces are we referring to when we talk about local climate action? And are we primarily looking for this scale to provide us with the resources to combat this otherwise global problem? Or does the local scale offer us a more tangible space with which to appreciate the impacts of climate change and thus re-think how this environmental crisis could (and should) be reshaping our relationship with our nearest neighbours? In Part II of the book, we focus our attention on the *spaces* of local climate action and draw together four important contributions to thinking about the question of 'where' local climate action comes from.

Connecting together three of the four chapters is the fascinating question of how the required reductions in global CO_2 emissions have now trickled down to the local scale and the institutional, practical and ethical challenges that this throws up. In Chapter 5, Dyson and Harvey-Scholes offer valuable insight into a phenomenon that swept the United Kingdom, and indeed the world, during 2019—the declaration of a climate emergency by local governments. Whilst the emergency rhetoric conveys a sense of real urgency to act, the authors assess the extent to which local authorities have displayed the requisite ambition in terms of follow-up action and how this fits with existing institutional capacities. Before even getting to the question of how such emission reduction strategies for the local scale can be developed, however, in Chapter 6, Russell and Christie offer fascinating insight into the challenges involved with simply establishing carbon emission baselines at the scale of local jurisdictions. Then, in Chapter 7, Harvey-Crawford and Creasy provide a candid and powerful set of reflections on the risks associated with granting special privilege to emission reductions as the single most important aim for climate action at the local level.

In addition to demonstrating the extent to which climate action is already well underway at the local scale, together, these chapters powerfully illustrate the interwoven and interdependent nature of the local, national and global scales. Questions of local enablement by national government; coordination of neighbouring geographies; and geographically distributed climate justice loom large. Offering a timely non-anglophone perspective, then, Chapter 8 from Haupt, Eckersely and Kern situates some of these debates not only in a different geographical context, but in a more historical register. Through their analysis, they illustrate how the otherwise 'ordinary' German cities of Gottingen and Remscheid have been actively cultivating a progressive approach to climate action since the early 1990s. The author's chapter serves as a poignant reminder that inspiration for how to deliver action on climate change need not merely come from those parts of the world deemed to be most pioneering by contemporary mainstream narratives.

THE AGENTS OF LOCAL CLIMATE ACTION

A core element of local climate governance rests on the different players that help shape it and those that will have to work alongside core climate policies in order to galvanise change and sustain ambition at the local level. In particular, questions such as whose role is it to drive local climate action? How are producers and users of scientific knowledge working to inform local action? And what innovative processes of engagement exist to ensure the voices and needs of different stakeholders are heard and incorporated into local climate action? The third, and final part of this book brings together four chapters that look at some of the key *agents* in local climate action and some of the innovative forms in which these agents interact with each other.

The four chapters in this final part provide a rich set of critical perspectives on core agents instrumental in leveraging and driving action on climate change. We begin with a broader perspective, with a focus on climate adaptation, where in Chapter 9 Guida and Howarth investigate how communication of climate knowledge to providers of climate change advice and support must consider the salience, credibility and legitimacy of such knowledge to align with the needs of different end users or agents of local climate action. In Chapter 10, Connell and Lane then explore how the private sector can benefit the work of city climate commissions by presenting a framework to better capture the diverse and varying set of competitive and economic motivators for private sector organisations to take action against climate change. In Chapter 11, Wells provides an account of citizen juries and assemblies on climate change to critically reflect on the extent to which they, as a process for democratic deliberation, credibly and legitimately engage the public on climate change. Finally, in Chapter 12, Robinson, Catney, Calver and Peacock explore the concept of the living lab in a university setting and their multiple roles to catalyse change locally through education, research and business engagement.

We see in this concluding part of the book, not only the wide range of agents involved in climate action, but their diversity and the benefits and complexities this brings to addressing the climate crisis through local action. This is explored through the lenses of climate adaptation and mitigation as well as innovative forms of agent collaboration, communication and engagement. In doing so, this part emphasises the core questions around how to most effectively represent and capture the needs, values

and capacities of local agents to contribute to a broader, yet effective, set
of impactful and inclusive climate actions.

Candice Howarth
Matthew Lane
Amanda Slevin

REFERENCES

Bulkeley, H., Marvin, S., Palgan, Y. V., McCormick, K., Breitfuss-Loidl, M.,
Mai, L., von Wirth, T., & Frantzeskaki, N. (2019). Urban living laborato-
ries: Conducting the experimental city? *European Urban and Regional Studies,
26*(4), 317–335.
Castán Broto, V. (2020). Climate change politics and the urban contexts of messy
governmentalities. *Territory, Politics, Governance, 9*(2), 241–258.
Freire, P. (1996 Edition). *Pedagogy of the oppressed.* (M. B. Ramos, Trans.).
Herder.
Galarraga, I., Gonzalez-Eguino, M., & Markandya, A. (2011). The role of
regional governments in climate change policy. *Environmental Policy and
Governance, 21*(3), 164–182.
Gouldson, A., Colenbrander, S., Sudmant, A., Papargyropoulou, E., Kerr, N.,
McAnulla, F., & Hall, S. (2016). Cities and climate change mitigation:
Economic opportunities and governance challenges in Asia. *Cities, 54,* 11–19.
HM Government. (2017). *UK climate change risk assessment 2017* [Online].
https://assets.publishing.service.gov.uk/government/uploads/system/upl
oads/attachment_data/file/584281/uk-climate-change-risk-assess-2017.pdf.
Accessed November 21, 2018.
Howarth, C., Barry, J., Fankhauser, S., Gouldson, A., Lock, K., Owen, A., &
Robins, N. (2021). *Trends in local climate action in the UK.* A report by the
Place-Based Climate Action Network (PCAN).
IPBES. (2019). *Global assessment report on biodiversity and ecosystem services of the
intergovernmental science-policy platform on biodiversity and ecosystem services*
(E. S. Brondizio, J. Settele, S. Díaz, & H. T. Ngo, Eds.). IPBES secretariat.
IPCC. (2018). Global Warming of 1.5°C. *An IPCC Special Report on the impacts
of global warming of 1.5°C above pre-industrial levels and related global green-
house gas emission pathways, in the context of strengthening the global response
to the threat of climate change, sustainable development, and efforts to eradicate
poverty* (V. Masson-Delmotte, P. Zhai, H.-O. Pörtner, D. Roberts, J. Skea, P.
R. Shukla, A. Pirani, W. Moufouma-Okia, C. Péan, R. Pidcock, S. Connors,
J. B. R. Matthews, Y. Chen, X. Zhou, M. I. Gomis, E. Lonnoy, T. Maycock,
M. Tignor, & T. Waterfield, Eds.).

Jordan, A., et al. (2018). *Governing climate change: Polycentricity in action?* Cambridge University Press.

Kivimaa, P., Hildén, M., Huitema, D., Jordan, A., & Newig, J. (2017). Experiments in climate governance—A systematic review of research on energy and built environment transitions. *Journal of Cleaner Production, 169*, 17–29.

PCAN. (2019). Available online https://pcancities.org.uk/what-local-climate-commission.

Slevin, A., Elliott, R., Graves, R., Petticrew, C., & Popoff, A. (2020). Lessons from Freire: Towards a pedagogy for socio-ecological transformation. *The Adult Learner: The Irish Journal of Adult and Community Education*, 73–95.

WMO. (2020). *WMO statement on the state of the global climate 2019.* (World Meteorological Organisation, WMO No. 1248).

Contents

Notes on Contributors

John Barry is a father, a recovering politician and Professor of Green Political Economy and Co-Director of the Centre for Sustainability, Equality and Climate Action at Queen's University Belfast. He is also co-chair of the Belfast Climate Commission.

Eugene Boadu is a Ph.D. Researcher in Management at Keele University. His research focuses on strategies for mitigating and adapting to climate change in sub-Saharan Africa, and the role played by multinational enterprise subsidiaries.

Philippa Calver is a Ph.D. Researcher with the Tyndall Centre for Climate Change Research at the University of Manchester and a Teaching Fellow at the University of Salford. Her research explores aspects of justice within the transition to a low carbon future.

Philip Catney is a Senior Lecturer in Politics at Keele University. He specialises in environmental and urban public policy and has published extensively in journals such as the *Journal of Environmental Policy and Planning*, *Environment and Planning C*, *Town Planning Review*, *Environmental Hazards*, *Journal of Environmental Management*, among others.

Rui Cepeda is a Ph.D. Researcher in Arts and Cultural Management, University of Manchester. His research focuses on the managerial and mediating processes used in the production of socially engaged art projects and its after-life, the documentation of those artistic projects. He has conducted ethnographic research examining the problems and conflicts affecting particular communities.

Debapriya Chakrabarti is a Ph.D. Researcher in Architecture and Urban Studies, University of Manchester. Her research investigates everyday practices and socio-spatial transformation of household-based crafts industry in an informal community due to shifting governance policies. She has conducted ethnographic and participatory research of studying lives and livelihoods of marginalised communities affected by disjointed infrastructures.

Ian Christie is Senior Lecturer in the social science of sustainable development in the University of Surrey's Centre for Environment and Sustainability. He is a member of the Surrey Climate Commission and a Fellow of WWF-UK. He is a co-investigator in the University's Centre for Understanding Sustainable Prosperity (CUSP, www.cusp.ac.uk).

Robert Connell while completing his M.A. International Business and Languages (German) at Heriot-Watt University, founded the university's Sustainability Society. Following this, Robert studied M.Sc. Carbon Management at the University of Edinburgh, where he focused on the topic of how the engage the private sector in taking greater action on climate change.

Alice Creasy is a Research Assistant at the Edinburgh Climate Change Institute focusing on issue of urban environmental governance. Alongside her work at the ECCI, she is a Policy and Membership Officer for the Local Government Information Unit and is a co-founder of the Embra Collective, an intersectional feminist collective that focuses on issues of climate change and sustainability. Alice has a first-class degree in Geography from the University of Glasgow and has a Masters in Environmental Sustainability (Distinction) from the University of Edinburgh.

Ami Crowther is a Ph.D. Researcher in Human Geography, School of Environment, Education and Development, University of Manchester. Her research focuses on low-carbon urban energy transitions, considering the actors, institutions and infrastructures associated with these processes and the interrelations between these entities.

James Dyson completed an M.Sc. in Environmental Policy and Regulation at LSE in September 2020, during which he investigated climate action in Manchester and gained a role as Research Assistant to Dr. Candice Howarth at ESRC-funded PCAN. Currently, James works as an environmental consultant for Ricardo and manages Editors for Impact CIC.

Peter Eckersley researches local climate governance, central-local government relations, public policy and accountability at the Leibniz Institute for Research on Society and Space in Germany, and at Nottingham Trent University, UK. Before entering academia, he spent ten years advising English municipalities on policy and management.

Kristen Guida is manager of the London Climate Change Partnership, after ten years at Climate South East, where she held the posts of Program Assistant and Coordinator/Director. In 2011, she co-founded Climate UK and chaired its Board of Directors until 2016. Her main interests are in supporting adaptive capacity, helping organisations formulate and share effective responses to the challenges presented by climate change, and bridging the gap between scientific evidence and action.

Rosanna Harvey-Crawford is a Project Officer at the Edinburgh Climate Change Institute. She is the coordinator for the Place-based Climate Action Network and the research group Research and Climate Action at the Local Level. She co-founded an intersectional feminist collective, Embra Collective, in 2020. She has a degree in French & Sociology from the University of Warwick and a Masters in Environment & Development from the University of Edinburgh.

Calum Harvey-Scholes is a Research Fellow in the Energy Policy Group at the University of Exeter specialising in local government, and community ownership and enterprise. He has a background in climate campaigning and is a trustee at *Stay Grounded* and the *UKYCC*.

Wolfgang Haupt has a background in Geography and holds a Ph.D. in Urban Studies from Gran Sasso Science Institute L'Aquila and Sant'Anna School of Advanced Studies, Pisa. His research focuses on local climate governance (mitigation and adaptation), transnational municipal networks, urban resilience and policy learning.

Teresa Hill has a background in the Community Theatre and Arts Sectors, working across the UK and Ireland for ten years before completing an M.Sc. in Leadership for Sustainable Development. She currently works for the Place-based Climate Action Network and the Centre for Sustainability, Equality and Climate Action at QUB.

Candice Howarth is a Senior Policy Fellow at the LSE Grantham Research Institute on Climate Change and the Environment and Co-Director of the Place-based Climate Action Network (PCAN). Her research interests focus on how the co-production of knowledge and science communication can be used to better inform decision-making in the context of climate resilience and sustainability challenges. She has an interdisciplinary background in climate policy, communication and pro-environmental behaviour with degrees in meteorology (B.Sc.), climate change (M.Sc.) and a Ph.D. in climate policy and pro-environmental behaviour.

Kristine Kern has a background in Public Administration, Economics and earned her Ph.D. in Political Science (Freie Universität Berlin). Kristine has worked and researched in Germany, the USA, Sweden, Finland and The Netherlands. Her current research interests concentrate on climate and energy governance, transnational municipal networks and sustainability transitions.

Matthew Lane is a Researcher in Sustainable Urban Governance at the University of Edinburgh. His research focuses on how city and regional governments are coping with an increased responsibility to act on crises of sustainability despite having limited legal, institutional, political and economic capacity to do so. He has undertaken fieldwork in the UK, Zambia, China and the USA.

Richard McLernon is Resilience Coordinator in Belfast City Council, managing programmes on climate and resilience, Belfast One Million Trees Programme and Belfast Urban Childhood Framework. Richard worked in community development in South and North Belfast, and with victims/survivors of the NI conflict, before joining Council to work in Community Safety.

Gustavo Montes de Oca is a writer and social entrepreneur currently developing a platform coop for community-owned shared electric mobility. He explores commoning solutions with Our Common Climate.

Ellie Murtagh is a Climate Adaptation Services Specialist at Sniffer, a Scottish sustainability charity. She works on the Adaptation Scotland programme, a Scottish government-funded initiative which provides advice and support to help organisations, businesses and communities prepare for and build resilience to the impacts of climate change. Her role involves supporting the public sector and benchmarking their adaptation progress, collating learning on place-based adaptation and climate finance.

Abderrahim Nekkache is a Ph.D. Researcher in Business and Management at Alliance Manchester Business School, the University of Manchester. His research focuses on digital transformation, innovation management and sociology of work.

Pauline O'Flynn is an organiser working in human rights and social and environmental justice in Belfast. She is currently involved in campaigns for sustainable neighbourhoods, mutual aid and community growing initiatives through her work with Participation and Practice of Rights (PPR) and Grow Community Gardens.

James Orr has an academic background in law, town planning and leadership. Currently, he is the Director for Friends of the Earth in Northern Ireland and a Climate Commissioner for the Belfast Climate Commission. In his previous career, he worked in local government and the Wildfowl and Wetlands Trust.

Adam Peacock is a Postdoctoral Research Associate within the Institute of Sustainable Futures at Keele University. His core research interests include user-centric design in Smart Local Energy Systems, consumer perceptions within energy transitions and changing rural-urban governance. He also has a background in Geographic Information Systems.

Zoe P. Robinson is Professor of Sustainability in Higher Education and Co-Director of the Institute for Sustainable Futures at Keele University. She is a Researcher, Educator and Practitioner in the field of sustainability science. Her current research focuses on user-centred and governance approaches to sustainability and the net-zero transition.

Erica Russell is a part-time Postdoctoral Researcher focusing on climate change and carbon emissions at sub-national level, climate governance, carbon embodied in supply chains—with a specific focus on the built environment and land use. Consultancy work: supporting new climate change business models, embodied carbon in the built environment and land use.

Amanda Slevin is an Environmental Sociologist with 20+ years experience in community development, adult and community education. Co-Director of QUB's Centre for Sustainability, Equality and Climate

Action, Amanda works with the Place-based Climate Action Network through which she co-founded Belfast Climate Commission and chairs its Community Climate Action Working Group.

Lucy Stone is a Climate Writer and Strategist working in philanthropy. She is a Director of a climate foundation and also on the board of a community energy organisation, and explores commoning solutions with Our Common Climate.

Lynda Sullivan is a writer and activist. She previously worked for human rights organisations in Ireland before spending 5 years in Peru accompanying Andean communities in their resistance against mega extractive projects. She is currently working on the issues of climate justice and extractivism with Friends of the Earth NI.

Bob Walley is a Lecturer in Community Leadership and Positive Environments Project Coordinator, University of Central Lancashire. He has managed local, national and international engagement projects concerning climate change communication and resilience for over 20 years, focusing on levels of empowerment, mobilisation or resistance. He is also a co-founder of the Manchester-based volunteer-led cooperative Envirolution.

Rebecca Wells graduated from her MSc Environment and Development at the London School of Economics in December 2020. Her Masters' dissertation research was a comparative analysis of the Oxford Citizens' Assembly and the Leeds Climate Change Citizens' Jury. Rebecca presented this research at the Place-based Climate Action Network's Climate Praxis Conference in September 2020. She now works in the UK Civil Service.

Paloma Yáñez Serrano is a Ph.D. Researcher in Social Anthropology with Visual Media, University of Manchester. She is an independent ethnographic filmmaker and social anthropologist interested in methods of adaptation humans develop to address changing environment, technology and political conflicts. Her research focuses on people's adaptation to industrial agriculture and changing landscape in southern Spain.

LIST OF FIGURES

LIST OF TABLES

PART I

Community and Place in Local Climate Praxis

CHAPTER 1

Local Climate Praxis in Practice: Community Climate Action in Belfast

Amanda Slevin, John Barry, Teresa Hill, James Orr,
Pauline O'Flynn, Lynda Sullivan, and Richard McLernon

Highlights Explores local climate praxis as transformative climate action with and in diverse communities across Belfast. Community climate action can tackle injustices, aid peace building and enable a just transition.

Keywords Climate breakdown · Multi-level climate action · Citizen participation · Community climate action · Climate praxis · Participatory action research · Northern Ireland

A. Slevin (✉) · J. Barry · T. Hill
School of History, Anthropology, Philosophy and Politics, Queen's University Belfast, Belfast, Northern Ireland, UK
e-mail: a.slevin@qub.ac.uk

J. Orr · L. Sullivan
Friends of the Earth Northern Ireland, Belfast, Northern Ireland, UK

P. O'Flynn
Participation and the Practice of Rights, Belfast, Northern Ireland, UK

© The Author(s) 2022
C. Howarth et al. (eds.), *Addressing the Climate Crisis*,
https://doi.org/10.1007/978-3-030-79739-3_1

As the COVID-19 pandemic forcefully disrupts our social world, it offers
a glimpse of large-scale social unrest, accelerated mortality and multi-
level inequalities inherent to the climate and ecological crises. It also
illustrates some salutatory lessons that could be 'read across' from how
states and communities have responded to the pandemic and they could
or should respond to the planetary emergency (Barry, 2020). Faced
with irrefutable scientific evidence of climate breakdown, many governments have declared a Climate and Ecological Emergency, including
the Northern Ireland Assembly (February 2020); however, Northern
Ireland (NI) is the only part of the UK without its own climate legislation. Offering hope that NI may soon develop effective climate policy,
recent collaborations between Climate Coalition Northern Ireland (NI's
largest civil society network for climate action), cross-party politicians and
legal experts culminated in introduction of NI's first Climate Change
Bill (Macauley, 2021). While emergency declarations and evolving policy
frameworks are central to societal shifts towards a sustainable future,
change is required across macro-, meso- and micro-levels of our social
world. A fundamental question we must consider is how do we enable
a genuinely inclusive, just transition to low carbon, healthier and fairer
communities and societies?

Evolving partnerships of public, private and third sector groups offer
insights into this crucial matter, leading the way in place-based climate
action, conceptually and practically, as we elucidate in this chapter on
community climate action co-written by members of the Community
Climate Action Working Group of Belfast Climate Commission.

PLACE-BASED CLIMATE ACTION IN BELFAST

With approximately 343,542 people, 18% of the population and almost
30% of all jobs in NI (BCC, 2020), Belfast city is home to NI's first
Climate Commission (the first such commission on the island of Ireland).
Formed in 2019, Belfast Climate Commission is one of three city-based
climate commissions and two theme-based platforms established as part
of the ESRC-funded, UK-wide Place-based Climate Action Network
(PCAN). PCAN brings together researchers and actors from the public,

R. McLernon
Belfast City Council, Belfast, Northern Ireland, UK

private and third sectors to collaborate on translating climate policy into action 'on the ground' to bring about transformative change. Co-chaired by Queen's University Belfast and Belfast City Council (BCC), Belfast Climate Commission develops robust evidence to inform place-based climate action in the region, in partnership with local stakeholders and PCAN partners. Involving Commissioners and Commission members, the Commission's four working groups illustrate its thematic orientation: Business and Finance; Community Climate Action; Just Transition; and Youth. The Commission is one strand of city-level climate action, complimenting and contributing to BCC-led endeavours like the *Belfast Resilience Strategy* and *One Million Trees* partnership project (BCC, 2020).

Amidst innovations in city-level policy and practice lie major challenges—we live in a high-carbon society in which macro-level systems can be reticent to change, thus leading to 'carbon lock-in'. At micro- and meso-levels, challenges include public understanding, democratic legitimacy, acceptance and engagement with essential multi-level climate actions; and further compounding these difficulties are limited state, business and civil society capacities to drive climate mitigation and adaptation within communities and wider society. The 'Carbon Roadmap' developed by Belfast Climate Commission and PCAN colleagues is a vital starting point for city-level GHG reduction targets, elucidating key sectors that require urgent attention—domestic housing (39% of emissions), transport (20%), public and commercial buildings (24%) and industry (18%) (Gouldson et al., 2020). Such research demonstrates the necessity of targeted climate action across society, from individuals and communities, through SMEs and industry, to local authorities and regional infrastructure. The Commission seeks to affect change across these various, interconnected levels because 'a resource efficient, low carbon and climate resilient city will not only be a better place to live, work and visit; it will also be a more prosperous and resilient place, better placed to respond to future economic and environmental shocks' (Belfast Climate Commission, 2019, p. 1). With a focus on 'bottom-up' approaches, the Commission's Community Climate Action Working Group (CCAWG) seeks to co-develop participative methodologies for climate action in and with communities across Belfast, as essential micro- and meso-level components of a just transition.

COMPLEXITY OF COMMUNITY
CLIMATE ACTION IN BELFAST

Globally, we see exciting examples of citizen involvement in the transition to a sustainable future, including community energy projects, tree planting for carbon sequestration, community gardens, eco-villages and social movements like Extinction Rebellion Youth Strikes for Climate and Transition Towns. Recognising the importance of citizen participation, key questions underpin the work of the CCAWG—effective climate action requires changes at all levels of society, how do we involve *all* citizens in community-level climate action? Especially working-class communities, young people and others beyond the 'usual suspects' one sees represented in standard city and state-based green policy consultation processes? That is, the engagement and participation (not simply passive consultation) of citizens beyond what are sometimes viewed as the 'urban, educated guilty middle classes'?

'Community' is often regarded a locus for meso-level citizen participation, yet how do we understand community for the purpose of climate action? This question is most pertinent in Northern Ireland where the legacy of ethno-nationalist conflict means 'community' can be used in an exclusionary sense to separate 'us' from 'them'. As Anderson (1991) articulates in his work on 'imagined communities', people can frame their identity in relation to national affiliation, whereby understanding of self is entwined with attachment to a nation-state, even when one can never know all members of such a large community. Due to 'The Troubles', as the conflict between NI's main social groups became known, challenges surround interpretations of 'community' and, at the risk of over-simplification and glossing over a long, painful history, a prevalent binary assessment means many people regard society as comprising two communities, Catholic and Protestant (BCC, 2020, p. 56). The national and religious identities of inhabitants are associated with NI's 'history of violent conflict' and some suggest Protestant communities largely identify as British whilst Catholic communities largely identify as Irish (Ramsey & Waterhouse-Bradley, 2018). Devastating outcomes of 'The Troubles' include the deaths of over 3,500 people (14 July 1969–31 December 2001) (Sutton, 2020) and over a third of people are estimated 'to have experienced a conflict-related traumatic event in their lifetime' (Griffin et al., 2019, p. 952). Although 'conflict is no longer the overriding risk factor for Belfast', its impact is revealed in continued division and segregation and the city is home to the highest number of interface areas in NI (BCC, 2020, p. 56). '97 security barriers and forms of defensive architecture' separate communities and such residential, physical, social and educational segregation contributes to division, low levels of trust, and can add to a city's vulnerability in times of crisis (ibid.).

To understand 'community' in NI, one must recognise the power of 'historic understanding and memory' (Deane, 1994) and how conflict-related division, trauma and hurt interact with deeply embedded socio-economic inequalities to produce negative outcomes for many citizens, and pose difficulties for transformative climate action. The term 'cross-community' has come to describe collaboration of people with differing affiliations along nationalist / religious lines; however, with inclusive, collective efforts to advance multi-level climate action, we can affect change '*a*cross-community', bringing together people from across society[1] to co-develop a healthier, sustainable and more peaceful society (Slevin, 2019). After all, 'climate breakdown doesn't care if we're Catholic, Protestant or Atheist' (ibid.), and, though space does not permit a fuller elaboration, there are connections between community peace-building, the riven places of Belfast and indeed 'peace' or sustainability with the planet. The work of Belfast Climate Commission is (implicitly) oriented towards the dynamic interplay of people, peace, planet and place—a divided city and people, facing the climate emergency.

LOCAL CLIMATE PRAXIS IN PRACTICE

Commitment to collaboration with communities is inherent to the Community Climate Action Working Group, which comprises members from public sector organisations, community, voluntary, environmental and activist groups. We share Ledwith's view that community is 'a complex system of interrelationships woven across social difference, diverse histories and cultures, and determined in the present by political and social trends' (2007, p. 32). Community can take various

---

[1] For example, categories specified in Section 75 of the *Northern Ireland Act* (1998): persons of different religious belief, political opinion, racial group, age, marital status or sexual orientation; men and women generally; persons with a disability and persons without; and persons with dependants and persons without (Equality Commission, 2010, p. 7).

forms including (1) territory or locality; (2) a communality of interest or an interest group; (3) a group sharing a common condition or problem (Wilmott, 1989, p. 2). Such framing allows us to consider climate action across Belfast's four main geographical areas (North, South, East, West), being cognisant of the city's unique social characteristics and diverse communities. In a city like Belfast, which bears the scars of prolonged conflict, participative approaches to community climate action may not emerge organically in a manner that transcends old divisions—meaning new forms of collaboration are necessary to go beyond dichotomous interpretations of community to create an inclusive community of communities committed to climate action.

Following the emergence of the coronavirus and subsequent lockdown, the CCAWG held its first meeting online in April 2020 and over two months, collaboratively articulated our overarching vision of co-creating 'transformative climate action with and in communities across Belfast'.[2] The CCAWG's attention to bottom-up approaches to climate action is important given intersections between climate injustices and socio-economic inequalities (HM Government, 2017; IPCC, 2018; Mendez, 2015; UN, 2019) and the necessity of participatory approaches to community engagement, beyond 'tokenistic consultations' (McNamara & Buggy, 2017, pp. 449–450). Furthermore, the highest proportion of properties at risk of flooding in NI are in deprived areas (27%) (BCC, 2020, p. 52) and low-income households are likely to have lower capacity and resources to adapt to the consequences of climate breakdown (HM Government, 2017, p. 10).

In the CCAWG's development phase, we considered principles and priorities to underpin our collective work and agreed to undertake research on community climate action to strengthen our evidence base and make informed decisions about future initiatives. We decided to undertake a participatory action research (PAR) initiative (ethical approval granted by Queen's University Belfast) as PAR is research concerned with change 'rather than simply understanding' (Dunne et al., 2005, p. 25) and our mixed methods research is underpinned by the explicit aim of gathering data to aid reflection and collaboratively enhance community climate action in Belfast. Influenced by Paulo Freire's articulation of praxis as 'reflection and action upon the world in order to transform it'

---

[2] For more information on the CCAWG: https://www.belfastclimate.org.uk/community-climate-action-working-group.

(Slevin et al., 2020), our approach to climate praxis aids knowledge co-production to co-create visions, ideas and knowledge as a basis for specific, contextualised solutions (Furuya, 2016).

In autumn 2020, we commenced an exploratory survey to gain insights into community climate action in Belfast and possibilities for future collaboration. The voluntary, opt-in nature of a formal online survey and its launch during the COVID-19 pandemic were not conducive to recruitment, particularly as many of our target group were either on furlough/ unemployed/ or balancing work and family commitments. Consequently, 25 people completed our survey; even with a limited sample, valuable findings emerged from the contributions of respondents who undertook paid and unpaid work with organisations whose reach spanned activities in one community to national and international groups, across domains such as climate activism, community development, environmental NGOs, international development and the public sector. Participants were located in North, South and East Belfast and the wider Belfast region, but there were no participants from West Belfast. Of those who participated in the research, 15 people identified as female, 10 as male and the age range spanned 18–66+ years (majority of participants were aged 35–54, $n = 13$).

Respondents shared diverse understandings of community climate action that ranged from local-level practical actions to reduce individual carbon footprints, through community-level activity as a momentum for broader social change, to critical multi-level analysis that encapsulated global-local interconnections. Interpretations included 'climate action that involves everyone in the community and that represents the diversity that exists in that community' (participant 2); 'encourage and enable all members of the community to engage in appropriate action to mitigate and ameliorate climate change' (participant 21); 'working together to save the planet' (participant 24). Participants shared their top priorities for community climate action, which included education, awareness-raising, reduction of GHG emissions, addressing inequality and influencing policymakers; they also outlined categories of climate action-related work they currently undertake and types of work they would like their organisation to undertake. The most popular forms of current activity were awareness raising ($n = 18$), information provision ($n = 12$), policy analysis ($n = 10$) and political lobbying ($n = 10$). In contrast, participants said they would like their organisations to undertake activities focused on community resilience initiatives ($n = 12$), educational programmes (formal [$n = 14$] and non-formal [$n = 11$]), political lobbying ($n = 10$).

Survey data has aided CCAWG identification of next steps and provided justification for new collaborations—18 of 25 participants said there is insufficient inter-organisational collaboration around climate change. Rationale for a lack of collaboration included resource constraints; groups working in isolation; communities not being invited to policy and strategic development initiatives; a lack of leadership from politicians and councillors with regard to climate action. Participants continued to outline characteristics, priorities and participants of potential Community Climate Action Networks and most participants said they would like to get involved with a network in their area (yes = 18, maybe = 6, no = 1).

MOVING FROM REFLECTION TO ACTION

As a new group, the Community Climate Action Working Group quickly coalesced around a shared goal of enhancing community climate action in and with communities, as a key element of place-based climate action. Our PAR initiative commenced with an exploratory survey to develop an evidence base to influence our activities; interviews and focus groups will be facilitated by the PAR PI (Slevin) to generate further data. Local climate praxis, as ongoing symbiosis of reflection and action, is inherent to our ongoing collaborations and a benefit of our PAR work thus far has been to raise the Commission's profile, culminating in new connections and ideas for community climate action. In tandem with planning community-level activities, the CCAWG has expanded to include some organisations who contributed to our PAR project, demonstrating early benefits of an initiative that will continue for some time. Even weaknesses within our exploratory survey offer valuable learning in terms of encouragement to undertake further engagement work with different communities.

As we strive to advance a just transition to a low-carbon, sustainable future, it is essential that we bring as many people as possible with us on the journey. In the case of Belfast, and Northern Ireland more broadly, societal inequalities and divisions pose significant challenges but in addressing those difficulties, we have opportunities to co-create inclusive, participative, place-specific approaches to enable capacity-building, collective action and impactful results across all levels of society.

References

Anderson, B. (1991). *Imagined communities: Reflections on the origin and spread of nationalism.* Verso.

Barry, J. (2020). *This what a real emergency looks like: what the response to Coronavirus can teach us about how we can and need to respond to the planetary emergency.* Available from: https://www.greenhousethinktank.org/uploads/4/8/3/2/48324387/this_is_what_a_real_emergency_looks_like_-_final_15-04-20.pdf

Belfast City Council (BCC). (2020). *Belfast Resilience Assessment.* Belfast: Belfast City Council.

Belfast Climate Commission. (2019). *Belfast Climate Commission Terms of Reference.* Belfast: Belfast Climate Commission. Available from: https://www.belfastclimate.org.uk/sites/default/files/BCCOM%20TOR%20vApril%202020.pdf

Deane, S. (1994). Living with our differences—Foreword. In E. Deane & C. Rittner (Eds.),*Beyond hate: Living with our differences.* Yes! Publications.

Dunne, M., Pryor, J., & Yates, P. (2005). *Becoming a researcher: A research companion for the social sciences.* Open University Press.

Equality Commission for Northern Ireland. (2010). *Section 75 of the Northern Ireland Act 1998: A Guide for Public Authorities.* Belfast: Equality Commission for Northern Ireland.

Furuya, S. (2016). *Sustainability praxis in community based renewable energy planning and development.* Aalborg Universitetsforlag (Ph.d.-serien for Det Teknisk-Naturvidenskabelige Fakultet). Aalborg Universitet.

Gouldson, A., Sudmant, A., Boyd, J, Williamson, R. F., Barry, J., & Slevin, A. (2020). *A net-zero carbon roadmap for Belfast.* Belfast Climate Commission/Place-based Climate Action Network.

Griffin, E., Bonner, B., Dillon, C. B., O'Hagan, D., & Corcoran, P. (2019). The association between self-harm and area-level characteristics in Northern Ireland: An ecological study. *The European Journal of Public Health, 29*(5), 948–953.

HM Government. (2017). *UK Climate Change Risk Assessment 2017* [Online]. [Accessed 21 November 2018]. Available from: https://assets.publishing.service.gov.uk/government/uploads/system/uploads/attachment_data/file/584281/uk-climate-change-risk-assess-2017.pdf

Intergovernmental Panel on Climate Change (IPCC). (2018). *Special report: Global warming of 1.5°C* [Online]. Accessed 24 July 2019. Available from: https://www.ipcc.ch/sr15/

Ledwith, M. (2007). *Community development—A critical approach*. The Policy Press.

Macauley, C. (2021). 'Historic' climate bill due before NI Assembly. *BBC* [Online]. Accessed 22 March 2021. Available from: https://www.bbc.com/news/uk-northern-ireland-56478081

McNamara, K. E., & Buggy, L. (2017). Community-based climate change adaptation: A review of academic literature. *Local Environment, 22*(4), 443–460.

Mendez, M. A. (2015). Assessing local climate action plans for public health co-benefits in environmental justice communities. *Local Environment, 20*(6), 637–663.

Ramsey, P., & Waterhouse-Bradley, B. (2018). 'Cultural policy in Northern Ireland: making cultural policy for a divided society. In V. Durrer, T. Miller, & D. O'Brien (Eds), *The Routledge handbook of global cultural policy*. Routledge.

Slevin, A. (2019, November 9). Climate breakdown: Impetus for place-based climate action and community development. Speech given at *Climate Jobs and A Just Transition: Responding to our climate emergency* conference, Queen's University Belfast.

Slevin, A., Elliott, R., Graves, R., Petticrew, C., & Popoff, A. (2020). 'Lessons from Freire: Towards a pedagogy for socio-ecological transformation'. *The Adult Learner: The Irish Journal of Adult and Community Education*, 73–95.

Sutton, M. (2020). Sutton index of deaths. *CAIN Web Service* [Online]. Accessed 18 January 2021. Available from: https://cain.ulster.ac.uk/sutton/

Wilmott, P. (1989). *Community initiatives: Patterns and prospects*. Policy Studies Institute.

Open Access This chapter is licensed under the terms of the Creative Commons Attribution 4.0 International License (http://creativecommons.org/licenses/by/4.0/), which permits use, sharing, adaptation, distribution and reproduction in any medium or format, as long as you give appropriate credit to the original author(s) and the source, provide a link to the Creative Commons license and indicate if changes were made.

The images or other third party material in this chapter are included in the chapter's Creative Commons license, unless indicated otherwise in a credit line to the material. If material is not included in the chapter's Creative Commons license and your intended use is not permitted by statutory regulation or exceeds the permitted use, you will need to obtain permission directly from the copyright holder.

CHAPTER 2

Putting the 'Place' in Place-Based Climate Action: Insights from Climate Adaptation Initiatives Across Scotland

Ellie Murtagh and Matthew Lane

Highlights Understanding and responding to the unique context and challenges of places is fundamental to the success of place-based adaptation projects.

Keywords Place · Climate adaptation · Partnerships · Scotland

INTRODUCTION

This chapter critically examines the concept of 'place' in relation to locally situated climate adaptation projects in order to better understand how this concept is being interpreted, negotiated and acted upon by practitioners. As evidenced by the contributions throughout this book, there is currently a groundswell of interest in focusing on 'place' as a scale

E. Murtagh
Sniffer/University of Strathclyde, Glasgow, Scotland, UK
e-mail: ellie@sniffer.org.uk

© The Author(s) 2022
C. Howarth et al. (eds.), *Addressing the Climate Crisis*,
https://doi.org/10.1007/978-3-030-79739-3_2

through which to deliver climate action projects and engender more climate literate individuals and communities. Following on from government discourse around place-*making,* the idea of place-*based* engagement with climate change and sustainability governance is gaining traction (PCAN, 2019; Vallance et al., 2019). Despite this, however, to date, little attempt has been made to understand how the concept of 'place' is understood by already existing climate action projects, and the role that place has played as an organising principle for coordinating action among co-located stakeholders.

To make an initial attempt at remedying this oversight, the chapter first briefly engages with the academic literature on the idea of place and place identity as well as emerging thoughts on their application to climate change adaptation specifically. Drawing on original qualitative research, the chapter briefly outlines the research project's methodological underpinning before discussing findings from interviews with lead practitioners from four adaptation initiatives in Scotland. We use these projects to ground some initial thoughts regarding the need to recognise the importance of place as a complex and value-laden mobiliser for collective climate action at a local scale.

Place in the Literature

The concept of 'place' is largely ineffable but is variously drawn upon to try and capture the complex interaction of agencies, objects and relationships which give meaning to particular locations. In turn, appreciation of the concept and the assigning of value to what it represents for individuals and communities can powerfully shape identity and attachment to particular places. The concept of place identity was first proposed by Proshansky (1978) who defined it as 'dimensions of self that define the individual's personal identity in relation to the physical environment by means of a complex pattern of conscious and unconscious ideas, beliefs, preferences, feelings, values, goals and behavioural tendencies and skills relevant to this environment' (Proshansky, 1978, p. 155). Paasi (1986) further developed the concept of place identity through distinguishing

M. Lane (✉)
School of Geosciences, University of Edinburgh, Edinburgh, Scotland, UK
e-mail: matthew.lane@ed.ac.uk

two separate facets of place identity, be that (i) the place identity of a place, which relates to how places are presented, promoted, identified or distinguished from other places through nature, culture, geography and history of that place; or (ii) people's place identity which pertains to how individuals exist within a place and their 'sense of place'. Place identity can thus influence behaviours and activities as well as individual and collective well-being.

More recently, it has been argued that place identity has particular relevance to climate change adaptation. Climate impacts will always be felt the strongest by individuals and communities most 'local' to those impacts. The more attached and embedded the relationships between those people and the places in question, the more devastating the impacts are likely to be. Fresque-Baxter and Armitage (2012) highlight that adaptation often focuses on the material assets required to build adaptive capacity, with less consideration to the nonmaterial or subjective facets of adaptation such as identity, belief or values. Value-based approaches can help explore what people value most about their everyday lives and how these social values are likely to be affected by environmental changes. It therefore seeks to redress the emphasis of adaptation planning on physical impacts by putting the lifestyle and wellbeing attributes that matter most to communities at the centre of adaptation planning (David et al., 2017, p. 167).

Place-based adaptation is interpreted within this chapter as actions which involve shaping, developing or enhancing a 'place' in response to current and projected climate change, whilst contributing to a broader context of change and social and ecological justice. Such actions, however, have had limited review or testing within empirical studies (Peng et al., 2020). With the aim of providing insight which can better inform future climate action projects seeking to pro-actively draw upon the idea of 'place' as an organising principle, in what follows we engage with four place-based adaptation initiatives in Scotland.

CASE STUDY AND METHODS

The Scottish policy landscape encourages place-based ways of working. The Scottish Government have adopted 'The Place Principle', which promotes a shared understanding of place and the need for a collaborative approach to a place's services and assets to improve outcomes for communities as well as contributing to the achievement of collective outcomes of Scotland's National Performance Framework (Scottish Government, 2019). The Community Empowerment Act also enables communities to have greater involvement in local decision-making by placing Community Planning Partnerships (CPPs) on a statutory footing to plan and deliver local outcomes.

Place-based working in Scotland is diverse in practice ranging from holistic explorations of place to focused initiatives on a specific issue, such as climate adaptation. The Climate Change (Scotland) Act 2009 requires a Scottish Climate Change Adaptation Programme (SCCAP) and also makes climate adaptation a statutory requirement for all Scottish public bodies, with mandatory reporting of progress. As a result of policy levers for place-based working and climate adaptation action alongside an increasing recognition of the importance of place, there are a growing number of place-based adaptation initiatives emerging.

In this chapter, we draw on semi-structured scoping interviews with the project leaders of four such initiatives, in order to understand the idea of 'place' as an organising concept. The projects explored were: Highland Adapts, Climate Ready Clyde, Climate Ready Aberdeenshire and Dundee Climate Action, defined in Table 2.1. The studied projects were all initiated by local government but are undertaken in partnership with wider stakeholders including community members, public and private organisations, in order to co-design local action plans. Sniffer, a Scottish sustainability charity, has been involved in some extent in all four projects, ranging from technical advice and capacity building through to providing secretariat support.

Interview questions explored (i) the interpretation and use of the concept of place within the various place-based adaptation projects, (ii) how variations in place have been identified and addressed through partnership projects and (iii) current appreciations for a sense of place among those participating in adaptation activities. Interview recordings were transcribed and thematic analysis drawn upon to identify emerging patterns within the data. Interview quotes are anonymised. To note, the views expressed in interviews reflect only those of the interviewees and are not meant to represent the way place is understood collectively by stakeholders involved with the respective projects. For the purposes of this short chapter, we are simply interested in how the concept of place is thought about and understood by those with important organising roles in 'place-based' climate action projects.

Table 2.1 Overview of place-based adaptation projects studies

Initiative	Description
Climate Ready Aberdeenshire	Climate Ready Aberdeenshire is a collaborative initiative to create Aberdeenshire's climate change adaptation and mitigation strategy. It brings together the views and expertise of a range of diverse stakeholders from communities, public, private and 3rd sector organisations, to set out how to work together to meet the challenges of a changing climate within Aberdeenshire, aiming to launch the Climate Ready Strategy in 2022
Climate Ready Clyde	Climate Ready Clyde is a cross-sector initiative funded by the Scottish Government and 15 member organisations to create a shared vision, strategy and action plan for an adapting Glasgow City Region
Dundee Climate Action	In June 2019, Dundee Council declared a Climate Emergency, recognising the serious and accelerating environmental, social and economic challenges faced by climate change. To respond to this challenge, a partnership Climate Action Plan has been prepared which has been the culmination of collaborative work, led by Dundee City Council and co-designed with public, private and community organisations, recognising that a concerted city-wide effort is required. The Plan includes four themes of Energy, Transport, Waste and Resilience
Highland Adapts	The Highland Adapts initiative brings organisations from across the Highland region together to develop a unique approach to adapting to climate change. Jointly resourced and rooted in a deep understanding of the needs and priorities of communities, the initiative will develop a strong evidence base setting out the climate risks and opportunities that are affecting the Highlands. This evidence will be used to develop a shared adaptation strategy and action plan which will embed action to adapt across organisational, community and sector plans, strategies and investments

FINDINGS FROM PRACTICE

The Concept of Place in Existing Place-Based Adaptation Projects

The four place-based adaptation projects explored here come from across Scotland and encompass rural, urban, coastal and highland environments. All project leads interviewed were cognisant of the diverse senses of 'place' that exist even *within* the areas they work. The feedback from practitioners presents an idea of 'place' interpreted as a multifaceted concept comprised of intersections of people, nature, built environment, culture, history and emotions in relation to a specific location. Place, in their opinion, supports the development of senses of identity, connections and feelings from interactions with or interpretations of the place; however, these may differ across individuals, communities and regions based on lived experience, demography or local geography.

The recognition of differing experiences and interpretations of 'place' was used to justify needing to think about place as a concept when undertaking situated climate action projects. All interviewees expressed strong feeling towards the need to work differently in each local place. Interviewees felt that a one-size-fits-all approach simply would not work, due to the specific challenges resulting from geographic, economic and demographic factors. Likewise, they felt that to explore and overcome these multiple perspectives, there was need to speak to people to develop a collaborative and joint view on what place means in the context of taking action against climate change. It was argued that this must be driven by and for the community, as those that live there know best what they want and need, as highlighted by one interviewee.

> A place-based approach helps because you're getting the views of individual people in that place...and the needs or challenges of each place are different, and you are actually taking that into account. It builds better engagement and relations to people in those places if you get to understand how they live and how they work, and not just paint them all with the same brush.

It was stressed by interviewees how both place and climate change are emotive subjects and need to be recognised and treated as such if they are to complement one another effectively. Interviewees felt that one cannot and should not take a non-emotional approach to adaptation.

> [You] can't not take an emotional response as climate change will impact lives, livelihoods, where people live, fundamentally for foreseeable future.

Focusing on a specific place acts as a mechanism to highlight the emotional, symbolic, spiritual and perceived intrinsic values of the environment. It is proposed by Adger et al. (2011) that climate change policy underemphasises the symbolic and psychological aspects of places and risk to them. This may be addressed through creating space and opportunity to share and discuss emotions.

> One of the key things about place is the intersect between culture and history...so many future challenges to adaptation are rooted in the past, whether that is from economic development to post-industrial legacy to issues with housing, stubborn social problems around health and vulnerability.

Place as an Organising Principle for Building Partnerships

All of the initiatives recognised the role of place, but not all had as of yet explored it methodically within the project. Activities so far were mainly high-level approaches ranging from community-wide surveys to inform the development of Highland Adapts; stakeholder workshops with participants across the public, private, third and community sectors to inform the Dundee Climate Action Plan; to the creation of a Theory of Change outlining a vision for a Climate Ready Glasgow City Region and necessary elements required to achieve it through the Climate Ready Clyde programme.

The ability to build understanding of different views and what place means collectively to frame wider discussion about the need to adapt was identified as to how the concept of place can be an enabler for climate change adaptation. Place was seen to make climate change impacts more tangible. A place-based approach can also enable identification of places or communities most 'at risk' of climate impacts and work to prioritise and support them.

Understandings of place, however, are arguably inseparable from broader multi-level influences of power, politics and policy that shape engagement with place. Macro-level decision-making about investments in place can lead to discord within areas that haven't received similar financial investment, in turn fostering a sense of disconnectedness to place which can hamper possibilities of 'place' as an enabler for climate action. An example of this was provided by a Scottish city experiencing massive regeneration in some areas, whilst other parts have persistently high levels of multiple deprivation.

Similarly, a noted barrier was tension in interpretations of place and the challenges in building and reaching consensus. The presence and activity of place-based adaptation partnerships were identified as mechanisms to overcome this challenge through participative processes with representative stakeholders. A multi-stakeholder approach can bring together stakeholders and community members to highlight important aspects of place, help co-design plans and shape activity. Place-based adaptation initiatives are seen to help create and anchor institutions in the process. In this respect, they can also help build and strengthen relationships. Interviewees felt, however, that whilst there may be different outcomes sought by each partner, the most important aspect of a joined-up approach is that the partnership is a collective listener to the needs and aspirations of the community and use their voice to shape overall response and enhance delivery.

Appreciating the Diverse (and Constantly Changing) Power of Place

Exploring, understanding and responding to the unique context and challenges of different places was seen as fundamental to the success of the place-based adaptation projects. In addition to this understanding, however, and in lieu of the fully integrated, joined-up approach to place-*making* reflected upon above, interviewees highlighted the need to appreciate the shifting relationships to place captured by any particular project scope. For example, interviewees felt that there is often a regional collective sense of identity, whilst simultaneously there are unique and bespoke senses of identity at the more local, community scale. For instance, within the area of Aberdeenshire, the identity of being from the 'Shire' and how this differentiated the lived experience of being from the 'city' (Aberdeen) was highlighted as providing a shared sense of identity, seeing a large region as one; whilst it was also noted that within the Highlands there are different senses of places across communities and even within a singular community, different sense of place exists.

In a similar vein, the onset of the COVID-19 pandemic during 2020 powerfully illustrates the way in which place attachment and identity can shift and evolve over time as well as from location to location. For example, interviewees suggest that the pandemic-associated lockdowns encouraged people to explore their local place in a lot more depth and potentially led to an increased appreciation for what's on the doorstep:

You are much more focused on your immediate environment obviously, because that's now where you go for entertainment, exercise, food, health...everything!

It's forced us to think quite deeply about what are the characteristics of a place that we value.

With regard to adaptation, these evolving relationships to place exhibit a very literal form of adaptation, further demonstrating its status as a constant process of change, rather than something that can simply be 'done' in a particular location at a particular point in time.

The pandemic has had significant benefit (to a point) in increasing people's awareness of their place and its associated advantages and limitations...[communities]have already been through a life changing emergency and can tap into feelings and experiences that they've gone through... to be prepared for future emergencies.

Conclusion: Lessons for 'Place-Based' Climate Action Projects

Places are shaped by services, assets, natural and economic resources, history, geography and environmental change as well as by the needs of the people who live there. Place-based adaptation projects offer a way to integrate ongoing activities, needs and local priorities to ensure a location is resilient to climate impacts as well as minimising the burden of diverse pressures which may be exacerbated under a changing climate. The motivation to further advance place-based adaptation is driven by the potential to create better places that retain their culture, identity and importance to those who live there and will be affected by the impacts of climate change.

Findings presented in this chapter highlight how an understanding of 'place' can enhance climate action projects. In accordance with the wider ambitions of this edited collection, the reflections from place-based adaptation project leads are presented here as potential learnings for others seeking to mobilise the concept of place as a tool for coordinating local action against climate change. However, given the context-specificity of the people-place-adaptation nexus discussed through this chapter, it would be counter-productive to provide specific recommendations for *how* place(s) should be engaged with by climate action projects. What the chapter *has* hoped to achieve, however, is a raising of awareness regarding what a greater appreciation for place may do to spur adaptation.

We argue here that this appreciation takes two contrasting yet equally important forms. Firstly, that place is a concept which has the potential to disrupt and undermine aspirations for climate action if practitioners are not acutely aware of the powerful (and diverse) ways in which people feel and experience attachment to certain locations. Secondly, that place can offer an organising principle which can be actively drawn upon to highlight already existing relationships between co-located stakeholders and between those stakeholders and the threats posed by climate change. As suggested by one interviewee, the success of climate adaptation projects will in large part be determined by local adaptive capacity. The complex nature of the relationship between adaptive capacity and 'place' thus calls for further research engagement. In the meantime, the more cognisant we are of the powerful and multifaceted ways in which particular *places* shape our collective understanding of climate change, the more likely it is that we will be able to take effective actions.

REFERENCES

Adger, W. N., Barnett, J., Iii, F. S. C., & Ellemor, H. (2011, May). This must be the place : Underrepresentation of identity and meaning in climate change decision-making. *Global Environmental Politics*, 1–26.

David, T., Graham, S., John, C., & Stephen, C. (2017, July). Advancing values-based approaches to climate change adaptation : A case study from Australia. *Environmental Science and Policy, 76,* 113–123. https://doi.org/10.1016/j.envsci.2017.06.014

Fresque-Baxter, J. A., & Armitage, D. (2012). Place identity and climate change adaptation: A synthesis and framework for understanding. *Wiley Interdisciplinary Reviews: Climate Change*, *3*(3), 251–266. https://doi.org/10.1002/wcc.164

Paasi, A. (1986). The institutionalization of regions: A theoretical framework for understanding the emergence of regions and the constitution of regional identity. *Fennia, 164*(1), 105–146. https://doi.org/10.11143/9052

PCAN. (2019). Available online: https://pcancities.org.uk/what-local-climate-commission

Peng, J., Strijker, D., & Wu, Q. (2020, March). Place identity: How far have we come in exploring its meanings? *Frontiers in Psychology, 11,* 1–19. https://doi.org/10.3389/fpsyg.2020.00294

Proshansky, H. (1978). The city and self-identity. *Environment and Behaviour*, *10*(2), 147–169.

Quinn, T., Lorenzoni, I., & Adger, W. N. (2015, January). Place attachment, identity, and adaptation. *The Adaptive Challenge of Climate Change*, 160–170. https://doi.org/10.1017/CBO9781139149389.010

Scottish Government. (2019). *Place principle: Introduction* [Online]. https://www.gov.scot/publications/place-principle-introduction/

Vallance, P., Tewdwr-Jones, M., & Kempton, L. (2019). Facilitating spaces for place-based leadership in centralized governance systems: The case of Newcastle City Futures. *Regional Studies, 53*(12), 1723–1733.

Open Access This chapter is licensed under the terms of the Creative Commons Attribution 4.0 International License (http://creativecommons.org/licenses/by/4.0/), which permits use, sharing, adaptation, distribution and reproduction in any medium or format, as long as you give appropriate credit to the original author(s) and the source, provide a link to the Creative Commons license and indicate if changes were made.

The images or other third party material in this chapter are included in the chapter's Creative Commons license, unless indicated otherwise in a credit line to the material. If material is not included in the chapter's Creative Commons license and your intended use is not permitted by statutory regulation or exceeds the permitted use, you will need to obtain permission directly from the copyright holder.

CHAPTER 3

A Commoners' Climate Movement

Lucy Stone, Gustavo Montes de Oca, and Ian Christie

Highlights The commoning approach to climate action collectively claims, creates and stewards the net-zero infrastructure. Commoning invites people to participate in the transition, to have a *stake*, not just a say, and shape their response.

Keywords Climate change · Commons · Ostrom · Community

COLLECTIVE ACTION ON CLIMATE

The climate transition is already under way, under the banner of 'net zero'—the aspiration to completely reduce (or offset) anthropogenic greenhouse gas emissions. The oil industry is in crisis whilst the renewables industry speeds ahead (IEA, 2020), agriculture faces transition,

L. Stone (✉) · G. Montes de Oca
Our Common Climate, London, UK
e-mail: lucy@ourcommonclimate.org

I. Christie
University of Surrey, Guildford, UK

© The Author(s) 2022
C. Howarth et al. (eds.), *Addressing the Climate Crisis*,
https://doi.org/10.1007/978-3-030-79739-3_3

whilst the car industry rapidly electrifies. These transitions and governmental action are welcome and necessary, but this top-down approach is still not generating effective measures fast enough to keep us within the level of global heating scientists deem 'safe' (UNEP, 2020).

These transitions largely exclude citizens, are not designed to avoid locking in existing inequalities and risk backlash over distributional consequences of climate transitions. All that could narrow the already tight political space in which elected representatives, governments and corporations operate (Willis, 2020). Many initiatives for citizens' engagement in climate policy measures have been launched (Capstick et al., 2020; Howarth et al., 2020). However, they have not been effectively connected to policymaking and they tend to treat people as individual agents/voters rather than as members of collective movements for change. We argue that many climate solutions are based on outdated models both of 'human nature' and of management of collective action problems. This constrains the 'possibility space' for action by overlooking the 'third pillar' of civil society—cooperative community-based action.

Our chapter draws from Elinor Ostrom's scholarship on managing commons (Ostrom, 1990) and a wider literature review, and our reflections as practitioners in the domains of community energy, agriculture and transport.

Commons and Commoning: What Are They?

Common land for grazing and woodland use is probably the most readily recognised form of commoning—ancient practices of managing shared access (though not necessarily via shared ownership). In the UK, legally defined 'common land' has declined from around 30% in the 1600s to 3% now (Shrubsole, 2019). In 2006, the UK Commons Act attempted to ensure that commons were owned *and* managed collectively, with new mechanisms for registering commons and to establish collective local governance mechanisms. In Scotland, where the 'Clearances' of common land took place through the nineteenth century, there are efforts to remedy this with legislation (the Land Reform Act 2003) and the subsequent movement for community land buy-outs, such as the 5,200 acres of Langholm Moor bought by the community.

Although Indigenous communities have legal ownership of just 10% of the world's land, they steward half of all collectively managed land. Their focus on community management protects large tracts of land from to land grabs, exploitation, deforestation and development, worsening the climate and biodiversity crises (Common Ground, 2016).

There is a wide array of models for collective self-governance of common resources available in the UK and globally in 2020. Here, we refer to 'Commoning' broadly as *collective self-governance of any common good resources by their co-producers and users*—an approach which is not just about the ownership model but about shared use, collective governance and sustainable stewardship over the long term. People who participate in these processes are 'commoners', and in their continuous evolution of rules and relationships to each other and the world around them, they demonstrate a richer capacity to collaborate than classic economics ascribes to people (Ostrom, 1990).

LOCKING OUT THE COMMONERS

Current approaches to climate change have inherited from the economic systems responsible for the crisis an outdated view of humans and a discredited understanding of collective action problems. The outdated view of persons as John Stuart Mill's *Homo Economicus*—individualistic, rational, self-interested maximisers—sees people as depleting the shared environment and collective resources because the individual benefits of extraction are higher than the costs, which are shared and may be displaced in time and space. Climate change solutions are therefore designed on the basis of government needing to adjust the market, to cost the externalities for rational individual choices (carbon trading and taxes, grants), rather than based on a deeper understanding of people as *HumanKind* (Bregman, 2020)—capable of altruism and collaboration. The outdated and incorrect understanding of shared resources, as *inevitably* misused—the so-called 'tragedy of the commons' (Hardin, 1968)—justifies a top-down, government-controlled approach.

Ostrom pointed out that climate policy has been constructed on the basis of this outdated theory of collective action, which is wrongly pessimistic about people's capacity for self-governance and overcoming of collective action problems (Ostrom, 2010). Ostrom won the 2009 Nobel Prize for Economics for discrediting the 'tragedy of the commons' framing of unsustainable resource use and showing, with detailed evidence that, *contra* Hardin, there are and have been many groups across cultures and socio-ecological conditions that successfully self-organise to manage common resources sustainably. The implicit model of the human, here, is based on reciprocity, shared interests and values (see, for example, Wright, 2000; Bregman, 2020).

Collective self-governance of common resources can in some circumstances be superior to top-down government—in the case of carbon, community forestry management sequesters more carbon than the government-run equivalent schemes (Chhatre & Agarawal, 2009).

CARBON MYOPIA: STATES SEE SYMPTOMS, COMMONERS SEE CAUSES

The effects of our model of abstraction, extraction, over-consumption and disposal are resulting in various ecosystem boundaries being threatened. There is mass biodiversity loss; soil degradation threatens farming collapse (FAO, 2020); microplastic pollution is ubiquitous (Cox et al., 2019); and air pollution is one of the leading causes of death worldwide (WHO, 2019). Excess carbon dioxide and other greenhouse gases, a by-product of this system of production and consumption, are a proxy for broader ecological devastation.

From the perspective of states and international bodies, the challenge of catastrophic ecosystem degradation has been reduced to a single measurable chemical, carbon dioxide, and the workings of a complex of nested problems reduced to a complicated system of accounting for this proxy. Managing for a single proxy, when the situation is far more complex, can create further problems. For example, in 2001, the UK government encouraged the sale of diesel fuel cars to reduce carbon intensity of UK transport. This singular focus—or carbon myopia—unintentionally intensified toxic air pollution.

The carbon-accounting view of climate change has been helpful in interpreting the science, to guide the parameters, and focus minds. The carbon focus does not guarantee a system of stewardship that remains within other ecological boundaries, let alone addresses inequality (Eisenstein, 2018).

POSSIBILITY SPACE

Recognising people in their full complexity and involving communities alongside markets and state increase the available range of possibilities for tackling environmental degradation. For example, community-based wind turbine schemes are able to secure community support where other forms of energy developments are blocked (Baxter et al., 2020). Local context, investment scale and local distribution of benefits to participating members reflect some of Ostrom's design principles for successful commons: Ostrom showed that when people are involved in common-pool resource decisions, they are more willing to accept consequences that might otherwise be seen as unacceptable 'sacrifices' (Ostrom, 1990).

In commoning, the one-dimensional picture of the human associated with neoliberal political economy is replaced by the more complex model Ostrom described. When people are directly involved in managing resources which they co-produce and use, the solutions that emerge benefit from the community's local knowledge. Management of common-pool resources on the basis of the design principles proposed by Ostrom (1990) secures the *stock* of resources and the *flow* of benefits required, and also generates *social trust* among the stakeholders. One commons will satisfy many needs of many stakeholders, resulting in solutions that improve local circumstances, but also propagate benefits into the environments they are nested in.

How Do We 'Common' the Climate?

It is clear that a stable climate, an ecosystem in balance, is a common good. However, the *atmosphere* is not a suitable candidate for management as a commons. According to Ostrom's design principles, the atmosphere would require clear group boundaries, but the atmosphere is nebulous and affects all humans, indeed all living beings.

Whilst the climate *symptom* of excess emissions needs urgent action, the commons approach focuses on the *cause*: the distorted relationships of people to planetary resources, assets and environments. The commoning approach is to collectively claim, create and steward the infrastructure of a net-zero world.

Land Use

Land use needs to be transformed in the coming decades—addressing deforestation, increasing tree planting, reducing agriculture impacts and restoring peatlands (CCC, 2020). Farmers face huge amounts of risk and uncertainty, which will only increase as the impacts of climate and other environmental crises grow. Community-managed or community-owned farms share the risks as well as the benefits with the farmers and share-holders sometimes provide labour as well. Community land trusts are created when communities come together to purchase land (or it is gifted) to develop and steward it for the benefit of the community over the long term. This may include community farms, or it could be to create housing developments rented at affordable rates or to create other community enterprises such as shared workplaces. A community land trust, as with community development trusts, is an institution collectively managed as well as owned, with benefits legally defined and safeguarded in perpetuity.

Creating land and other local assets in collective, local stewardship yields many benefits, direct and indirect (Capital Economics, 2020). Community land trusts have rapidly grown in the UK, mostly as a way to provide affordable housing but also to preserve local assets facing decline, such as pubs, post offices and shops. There are nearly 300 such land trusts, half of which have been established in the last 2 years alone. The developments are designed with the community's close involvement and permission, and freed from the pure profit motive, sustainability and affordability take priority. As the UK meets the shortage of housing whilst addressing climate change, community land trusts are a positive way to make transitions towards sustainability.

CLEAN ENERGY COMMONS

Renewable energy will be a dominant feature of the energy landscape, as well as of the landscape itself. Community energy systems are a commons managed in line with Ostrom's design principles (Melville et al., 2017, 2018). Community energy organisations raise funds locally from share-holder members for local renewable energy projects. Local knowledge and relationships make the development of the projects possible. Typically, they offer returns to shareholders as well as amassing community funds for improving localities in consultation with members. Community energy projects generate social, environmental and economic benefits for diverse stakeholders (Armstrong, 2015; Melville et al., 2017; Robinson & Stephen, 2020). For example, schools and community centres benefit from discounted electricity rates. In addition, many community energy organisations allocate funds to support fuel poverty through advice lines or installing energy efficiency measures (Armstrong, 2015).

Whilst generation capacity is currently not massive at around 500 MW, there are 286 community energy organisations in England alone, and this is growing despite a hostile policy context (Community Energy England, 2020). Many of these projects are in urban areas, using rooftops. The size of projects appears to be growing too as the sector matures and the technology costs drop—a project in Devon plans to install a distributed 1GW of electricity. This is not, as Ostrom would emphasise, a panacea. Commons models for community energy need careful design and governance to avoid exacerbating problems of mistrust, exclusion and inequality (ibid.). However, the evidence indicates great scope for policies that treat sustainable energy systems as commons and harness the capabilities of citizens and local communities alongside the private sector and national state (Webb et al., 2021).

A Climate Commons Narrative

The prevailing top-down climate policy narrative is not translating high public concern into the urgently needed political mandate for greater action. This narrative implies that an overwhelming, abstract and distant issue is being resolved at a central level by politicians and bureaucrats (Marshall, 2015). People receive the message of the urgency only to be told it is being managed by governments or only to be urged to make individual behavioural changes, which can feel tokenistic (Marshall, 2015), or simply to campaign or vote or deliberate about changes to be implemented by experts.

Climate narratives are falling into the trap of communicating climate as a single issue, rather than presenting it as the symptom of the deeper and bigger problem of unsustainable resource use that needs to be addressed. The narrative that we are separate from nature—which serves as a 'free' resource which we must manage, even when sustainably, for our use—is fuelling the destruction of the natural world (see, for example, White, 1967; Plumwood, 2001). It is not even true to the realities of our biology and ecological situatedness (Haraway, 2016)—we live in socio-ecological interdependence with a multitude of species from gut microbes to the pollinators who provide our food. Commoning approaches recognise our interdependence—with each other and with the land. (The origin of 'common land' in Gaelic, actually meant 'people together as one with the land' [Menzies, 2014].)

Commoning invites people to participate in the transition, to have a *stake*, not just a say, and to shape their own neighbourhood's response. It is a creative invitation to exercise more agency even than the 'citizen control' over planning that Sherry Arnstein (1969) called for. This space of action—the local, community level—produces tangible, meaningful results and multiple benefits (Kaye, 2020). This in turn can then create a political mandate for further change, whilst increasing local resilience.

The commoning narrative acknowledges that the climate crisis is a big problem, but invites people to address it jointly, not in isolation, and at multiple scales, not just top-down. It breaks the problem down and invites bottom-up community action, and multi-level partnerships. It is clear there will be upheavals in our patterns of production and consumption, but commoning calls for people to be active in bringing about the change rather than simply be the beneficiaries or unfortunate losers from it.

This sense of control and agency has been shown to be important factors in determining mental and physical health, community empowerment and social cohesion. By empowering communities to develop greater altruism and social support will help ensure communities are not just surviving but thriving as we mitigate and navigate climate disruption.

One challenge will be to ensure commoning doesn't create 'lifeboatism'—shoring up my community resilience whilst other communities face growing crises. Increasingly, though, there are options for self-organising groups to provide and procure from others, strengthening demand for each other's services. Can we one day connect with other commoners to procure goods from a community-owned manufacturer, with materials from miners' co-ops, powered by community energy and insured by a cooperative underwriter? Bringing this vision into climate spaces might help prevent the perpetuation of existing inequalities as we build the new green economies. In order to meet the climate crisis, we need to rapidly engage all of society; commoning might be the way we do this.

REFERENCES

Armstrong, H. (2015). *Local energy in an age of austerity*. National Endowment for Science, Technology and the Arts.
Arnstein, S. (1969). A ladder of citizen participation. *Journal of the American Institute of Planners, 35*(4), 216–224.
Baxter et al. (2020). *Scale, history and justice in community wind energy: An empirical review* (https://www.exeter.ac.uk).
Bregman, R. (2020). *Humankind*. Little, Brown and Company.

Capital Economics. (2020). *Housing by the community, for the community.* Community Land Trust UK.

Capstick, S., Demski, C., Cherry, C., Verfuerth, C., & Steentjes K. (2020). *Climate change citizens' assemblies* (CAST briefing paper 03). CAST, University of Cardiff.

Chhatre, A., & Agrawal, A. (2009). Trade-offs and synergies between carbon storage and livelihood benefits from forest commons. *Proceedings of the National Academy of Sciences of the United States of America, 106,* 17667–17670. https://doi.org/10.1073/pnas.0905308106.

Committee on Climate Change. (2020). *Land use: Policies for a net zero UK.* Committee on Climate Change.

Cox, K., Covernton, G., Davies, H., Dower, J., Juanes, F., & Dudas, S. (2019). Human consumption of microplastics. *Environmental Science and Technology Journal, 53*(12), 7068–7074.

Eisenstein, C. (2018). *Climate; A new story.* North Atlantic Book.

Food and Agriculture Organisation. (2020). *Status of the world's soil health, food and agriculture organisation of the United Nations.*

Haraway, D. (2016). *Staying with the trouble: Making Kin in the Chthulucene.* Duke University Press.

Hardin, G. (1968). The Tragedy of the Commons. *Science, 162,* 1243–1248.

Hoffman, M. (2013). Why community ownership? Understanding land reform in Scotland. *Land Use Policy, 31,* 289–297.

Howarth, C., Bryant, P., Corner, A., et al. (2020). Building a social mandate for climate action: Lessons from COVID-19. *Environmental and Resource Economics, 76*(4), 1107–1115.

IEA. (2020). *World energy outlook 2020.* Paris: IEA. https://www.iea.org/reports/world-energyoutlook- 2020.

Kaye, S. (2020). *Think big, act small: Elinor Ostrom's radical vision for community power.* New Local.

Marshall, G. (2015). *Don't even think about it; why our brains are wired to ignore climate change.* Bloomsbury.

Melville, E., Burningham, K. A., Christie, I., & Smallwood, B. (2018, April–June). Equality in local energy commons: A UK case study of community and municipal energy. *Rassegna Italiana di Sociologia, LVIX*(2).

Melville, E., Christie, I., Burningham, K. A., Way, C., & Hampshire, P. (2017). *The electric commons: a qualitative study of community accountability'. Energy Policy, 106.*

Menzies, H. (2014). *Reclaiming the commons for the common good.* New Society Publishers.

Ostrom, E. (1990). *Governing the commons; the evolution of institutions for collective action.* Cambridge University Press.

Ostrom, E. (2010). Polycentric systems for coping with collective action and global environmental change. *Global Environmental Change, 20*(4).

Oxfam, International Land Coalition. (2016). *Common ground: Securing land rights and safeguarding the earth*. Rights and Resources Initiative. Oxfam.

Plumwood, V. (2001/2002). *Environmental culture: The ecological crisis of reason*. Routledge.

Robinson, S., & Stephen, D. (2020). *Community energy 2020: The state of the sector*. Community Energy England.

Rose, C. (2011). Ostrom and the lawyers: The impact of governing the commons on the American legal academy. *International Journal of the Commons, 5*(1), 28–49.

Shrubsole, G. (2019). *Who owns England?* HarperCollins.

United Nations Environment Programme. (2020). *Emissions Gap Report*. United Nations Environment Programme.

Webb, .J, Stone, L., Murphy, L., & Hunter, J. (2021). *The climate commons: How communities can thrive in a climate changing world*. The Institute for Public Policy Research.

World Heath Organisation. (2019). Website. https://ourworldindata.org/air-pollution.

World Health Organisation. (2021). *Air pollution (who.int)* Website. Accessed 4 January 2021.

Willis, R. (2020). *Too Hot to Handle? The democratic challenge of climate change*. Bristol University Press.

Wight, L. (1967, March 10). The historical roots of our ecologic crisis. *Science, 155*(3767), 1203–1207.

Wright, R. (2000). *Nonzero: History, evolution & human cooperation*. Little, Brown and Company.

Open Access This chapter is licensed under the terms of the Creative Commons Attribution 4.0 International License (http://creativecommons.org/licenses/by/4.0/), which permits use, sharing, adaptation, distribution and reproduction in any medium or format, as long as you give appropriate credit to the original author(s) and the source, provide a link to the Creative Commons license and indicate if changes were made.

The images or other third party material in this chapter are included in the chapter's Creative Commons license, unless indicated otherwise in a credit line to the material. If material is not included in the chapter's Creative Commons license and your intended use is not permitted by statutory regulation or exceeds the permitted use, you will need to obtain permission directly from the copyright holder.

CHAPTER 4

The Envirolution Revolution: Raising Awareness of Climate Change Creatively Through Free and Accessible Community Engagement Festivals

Bob Walley, Ami Crowther, Paloma Yáñez Serrano, Abderrahim Nekkache, Rui Cepeda, Debapriya Chakrabarti, and Eugene Boadu

Highlights The use of Freirean praxis and pedagogy is a powerful lens to use in understanding, and advancing, impactful place-based climate action. Using non-judgemental and holistic approaches, free community festivals have the capacity to engage large sections of society and therefore have the potential to mobilise shifts towards societal change.

B. Walley (✉)
University of Central Lancashire, Preston, UK
e-mail: rvwalley1@uclan.ac.uk

A. Crowther · D. Chakrabarti
School of Environment Education and Development, University of Manchester, Manchester, UK

P. Yáñez Serrano
School of Social Sciences, University of Manchester, Manchester, UK

© The Author(s) 2022
C. Howarth et al. (eds.), *Addressing the Climate Crisis*,
https://doi.org/10.1007/978-3-030-79739-3_4

Keywords Climate emergency · Community engagement ·
Environmental education · Festivals · Critical consciousness

INTRODUCTION

Envirolution is a Manchester-based volunteer-led cooperative which
organises community engagement events concerning climate change. The
group creates spaces where a praxis of learning and reflection can take
place in a holistic, accessible and relevant way, aiming to engage those
individuals who are not currently engaging with climate change as a
subject. Using the teachings of Paulo Freire and the praxis model, Envirolution explores and develops creative ways to engage and empower
people, utilising this reflexive process to inspire participants to take positive actions. At events Envirolution has organised or been involved with,
the group has interacted with 189,400 attendees across the UK, involved
785 volunteers and provided a platform for 1142 educators, community
organisations, performers and artists. At a local level, Envirolution aims to
play a key role in helping Manchester City Council achieve goals set out in
its Climate Change Action Plan 2020–2025, which includes supporting
and influencing the city to reduce emissions by at least 50% by 2025.
In a wider context, Envirolution aims to raise widespread awareness and
engagement with the climate emergency and mobilise transitions towards
a more sustainable future. Despite engaging significantly high numbers
of participants since it began, the project's impact has only now been
properly evaluated. Using data from a University of Manchester research

A. Nekkache
Alliance Manchester Business School, University of Manchester, Manchester,
UK

R. Cepeda
School of Arts, Languages and Cultures, University of Manchester, Manchester,
UK

E. Boadu
Keele Business School, Keele University, Keele, UK

team's project analysis, this chapter shows how effective this method of engagement can be and what can be learnt from the Envirolution project.

METHODOLOGY

To evaluate how effective Envirolution events have been over the past ten years, the project analysis took a multi-methods approach. This included a scoping literature review, analysis of quantitative data on attendance and participation from previous Envirolution events, and a targeted online questionnaire survey.

In the associated literature review, existing understandings of community engagement and climate change action related to the Envirolution model were consolidated to situate the research and provide context for the analysis. The quantitative data collected by Envirolution volunteers at previous events were analysed and trends were identified. A set of questionnaires were used to collect information on individuals' experiences of Envirolution events, and to understand and analyse the impact that engaging with Envirolution had on understanding of climate change and behavioural responses. A total of 40 questionnaires were obtained from: 15 general public attendees, 13 stallholders or workshop providers, 10 Envirolution volunteers and 2 workshop leaders or speakers. Responses were analysed thematically, drawing out key phrases and concepts.

THE ENVIROLUTION MODEL

Praxis is described by Paulo Freire as 'reflection and action upon the world to transform it' (Freire, 1996). Throughout his work as an educator, Freire stimulated praxis through informed action and critical pedagogy as a means to liberate powerless communities which he referred to broadly as the 'oppressed'. Reflective praxis motivates participants to become critical subjects, aware of their own contradictions (Freire, 1968). Freire suggests that the role of the educator is to help others in their own transformation, 'placing them in a consciously critical position before their problems' (Freire, 1968, p. 50). The aims of Envirolution are guided by Freire's work and events are created with the ethos and understanding that considered and careful reflection can lead to positive actions.

As Freire emphasised, 'the insistence that the oppressed engage in reflection on their concrete situation is not a call to armchair revolution. On the contrary, reflection – true reflection – leads to action' (Freire,

1996, p. 41). Praxis in this sense goes beyond activism in that it pursues actions that emerge from critical reflection. With this philosophy, Envirolution brings together individuals and organisations who ask themselves the same question: what can I possibly do about climate change? The events provide a platform to engage those who may have similar questions, or due to the incomprehensible scale of the problem, have not considered a response at all.

Envirolution primarily engages people from the local areas of Fallowfield, Rusholme, Longsight and Moss Side in Manchester, representing a multi-cultural demographic of various ages, backgrounds and circumstances. Well over half of the attendees are in the park on the day of the festival simply because it is their local green space, which then gives them the opportunity to access a free event. All attendees are introduced to facilitators who can provide opportunities for reflection on climate change, offering relevant and considered positive responses. This allows participants to choose and decide their actions through critical reflection, critical participation and collective action. Envirolution festivals aim to encourage this critical consciousness process in a free and non-judgemental environment, providing a space for praxis and exploration of possible 'transformation of the world' (Freire, 1996, p.106). Open free community spaces like these could prove crucial for exploration of the multi-level changes needed to avoid climate collapse, which depends on 'the public's willingness to accept, support, and actively engage' with the required socio-economic, cultural, political and structural shifts (Geiger & Swim, 2016, p.79). The environmental festivals Envirolution volunteers organise work on the basis that 'people's individual issues lead to local projects; local projects link with others, elsewhere, to form networks and alliances; those alliances lead to movements that provide a real collective possibility for change' (Ledwith, 2007, p. 609).

The Envirolution model also acknowledges how crucial emotional responses are to understanding people's perceptions or levels of engagement with the climate emergency. Climate change educator and researcher Blanche Verlie identified 'anxiety, frustration, feeling overwhelmed, guilt and grief' as natural responses to such an unprecedented global phenomenon as climate change (Verlie, 2019, p. 751). Engaging people about impending global social and ecological destruction by highlighting the growing consensus that 'climate change is the result of human lifestyle and behaviour' (Roeser, 2012) can ignite many emotional responses, not all of them productive. By concentrating on effective adaptation, 'hope' or 'positive engagement' can also be created through the forming of participatory groups (Verlie, 2019 p.751). Critical consciousness, in this sense, helps participants see the world and their place within it, acknowledging their emotional responses as a crucial element in the process of praxis and enabling progressive steps forward.

Findings and Discussion

Engagement with Envirolution events has steadily increased over the past 10 years. At the inaugural Envirolution festival in 2010, there were 400 attendees, 52 activity providers and 22 volunteers. This increased to 5000 attendees, 89 activity providers and 52 volunteers in 2019. Those that engage with Envirolution events consider them to successfully create 'a fun and friendly environment' with educational and awareness activities that inspire them to 'make positive changes for a sustainable future' (Envirolution, 2021). When focusing on the impact that Envirolution has had on individuals that engage with the organisation, 4 key areas of impact were identified through analysis of the survey undertaken by the 40 participants.

Learning and Awareness

> Positive impacts only happen when people change their attitudes. (Volunteer 1)

Questionnaire responses found Envirolution to be 'educational', 'informative' and 'inspiring'. Envirolution events aim to try to appeal to as many people as possible, so the solar-powered music stages play a range of world music and act as a hook to attract people from all cultures, ages and backgrounds. Once at the event, attendees can learn something new and see what they can get involved with locally. Attendees can listen to 'inspirational speakers' (Attendee 1) or 'immerse themselves in environmental activities' (Stallholder 7). The research also found that a considerable number of participants shared what they learnt with others (54% of attendees and 84% of volunteers).

If these kinds of events happened in parks around the country, I believe the population would be a lot more active in responding to the climate emergency. (Volunteer 4)

Behavioural Change

Even small changes can have a significant impact. (Attendee 1)

83% of stallholders who participated in the research believe that their activities support positive changes in people's daily activities and behaviours. 90% of volunteers and 69% of visitors state that engaging with Envirolution made them more environmentally conscious, leading them to make changes to their daily activities. 5 of the 40 respondents confirmed that thanks to Envirolution, they have become climate change activists who actively campaign and engage with community groups, regularly sharing their knowledge with others.

The groundswell of support from the population, especially a younger generation fighting for their future, gives hope. (Stallholder 7)

Collaboration

Knowing I'm not the only one with this feeling. (Volunteer 1)

After Envirolution, many attendees have collaborated on environmental projects and learnt about other groups they can get involved with. Also, taking part in Envirolution events has boosted participants' confidence, enabling them 'to volunteer for environmental protection groups' (Volunteer 3). Inference from research responses demonstrates that Envirolution holds great potential for collaboration between multiple organisations and groups. This sharing of learning with others is inherent to the reflexivity and action with Freire's articulation of praxis. Effectively engaging large numbers of people in this manner can be of great value for local councils or national governments as part of action plans towards climate change mitigation and resilience strategies. So why are similar events not happening all over the UK? Or indeed the world? 'I don't know of another grassroots organisation like Envirolution. I would love to see initiatives like this in other cities and towns across the UK' (Stall-holder 1). As a volunteer-led organisation, Envirolution organisers only have so much capacity and putting on large engagement events provides the team with plenty to do already. But this raises interesting questions about how this method of engagement could be improved or expanded in collaboration with strategic stakeholders and policymakers.

> We are no longer individual persons, each struggling alone to survive the whims of nature. Increasingly, we share the same fate. (Stallholder 12)

COMMUNITY ENGAGEMENT

> I fear it is too little too late for this world... I hope the majority realise we have to act fast and not leave the responsibility to a select few, it's everyone's problem! (Volunteer 9)

80% of attendees surveyed stated that Envirolution 'heightens aware-ness of environmental issues and promotes community spirit'. Of the stallholders surveyed, 75% said they were able to engage with a wider audience by participating in Envirolution. However, as stated the capacity of volunteer-led groups like Envirolution to be able to engage more people is limited. Project volunteers acknowledge the need to engage more participants from local religious centres, schools and other commu-nity groups in the lead up to events. With more support from local councils and other stakeholders increasing capacity, significantly greater levels of impact with a wider audience could be achieved.

CONCLUSION AND RECOMMENDATIONS

> My fears are that we are too late to repair a lot of the damage our way of life has caused to our home world, and that history will not look kindly on this time. My hopes are that enough people respond to this emergency before much more damage can be done, and this results in us creating a more balanced and fairer society for all which values the natural world. (Volunteer 4)

The research of these methods of public engagement shows the abundant potential they hold in creating greater awareness of climate change and a 'growing understanding that social and environmental issues are inter-connected and require a whole society approach' (Head, 2005). Groups like Envirolution can undoubtedly help this transformational learning process. The effective evaluation of the project was both timely and crucial for better understanding of why people engage with Envirolution, and therefore climate change, identifying motives and perceptions that can inform other approaches to climate change education and engagement. This research certainly shows Envirolution and similar events can act as gateways for people to try and make sense of the climate emergency that is upon us and find a proactive way of responding. Project volunteers hope that by engaging individuals to become part of other groups and organisations, participants can begin to understand the necessity of collab-orative responses and the valuable roles they can play as part of a global movement at a local level.

However, as can happen all too often in climate change education, collective action for change is not followed through to its greatest poten-tial, and practice remains contextualised in the immediate, local and specific without making critical connections with the structural roots of what Freire would call the 'oppression', from which inequalities emanate. This can result in a fixation on symptoms, leaving the root causes free to perpetuate oppressions. 'We debate responsibilities over rights, as the responsibility for an unjust system is turned upon the victims of that injustice and the radical discourse of social justice is subtly absorbed and distorted into a rhetoric of self-help' (Ledwith, 2007). This can be seen in government carbon reduction campaigns across the world, focused on directing the population to reduce their personal carbon footprint whilst still buying into the global market and only being given mainstream retail, food and transport options that remain heavily carbon intensive.

We need to situate our local practices within the bigger political picture to identify the way that structural oppressions get acted out in local contexts. A participatory process, one based on true democracy, aims to give autonomy and power to the voices of subordinated groups, accepting that there are many truths, rather than one universal truth or answer. As poorer countries in the Global South hold the least respon-sibility for accelerating climate change but already experience the worst consequences, this aim is now more important than ever. Freirean praxis informed approaches can elevate the diversity of human experience over the imperative of economic 'progress', locating social and environmental justice at its heart, whilst mobilising collective action for social change as its outcome. By combining praxis and critical consciousness focused social movements like Envirolution with wider political climate change mitigation and adaption targets, it becomes possible to create combined and informed responses to the climate emergency, which can be communicated and facilitated with large groups of the population. The urgency represented by possible socio-ecological collapse calls for more informed and combined responses between individuals and groups at all levels of society, promoting critical consciousness, and a culture of praxis and reflection towards positive action.

The evaluation report findings and more information about Envirolution can be found on the website at: www.envirolution.org.uk.

REFERENCES

Bowen, F., et al. (2010). When suits meet roots: The antecedents and consequences of community engagement strategy. *Journal of Business Ethics, 95*, 297–318.

Envirolution and University of Manchester Collaboration Labs Research Project. (2021). *Envirolution Impact Report*. Available at: http://envirolution.org.uk/. Accessed 9 January 2021.

Freire, P. (1968). *Pedagogy of freedom: Ethics, democracy and civic courage.* Rowman and Littlefield Publishers, Inc.

Freire, P. (1996 Edition). *Pedagogy of the oppressed* (M. Bergman Ramos, Trans.). Herder.

Geiger, N., & Swim, J. K. (2016). Climate of silence: Pluralistic ignorance as a barrier to climate change discussion. *Journal of Environmental Psychology, 47*, 79–90.

Head, B. W. (2005). Community engagement—Explanations, limits and impacts. In D. Gardiner & K. Scott (Eds.), *Proceeding of international conference on engaging communities*. Queensland Department of Main Roads.

Ledwith, M. (2007). On being critical: Uniting theory and practice through emancipatory action research. *Educational Action Research, 15*(4), 597–611.

Lee, S. Y., Petrick, J. F., & Crompton, J. (2007). The roles of quality and intermediary constructs in determining festival attendees' behavioural intention. *Journal of Travel Research, 45*(4), 10–14.

Roeser, S. (2012). Risk communication, public engagement, and climate change: A role for emotions. *Risk Analysis, 32*(4), 1033–1040.

Slevin, A., et al. (2020). Lessons from Freire: Towards a pedagogy for socio-ecological transformation. *The Adult Learner*, pp. 73–95.

Verlie, B. (2019). Bearing worlds: Learning to live-with climate change. *Environmental Education Research, 25*(5), 751–766.

Open Access This chapter is licensed under the terms of the Creative Commons Attribution 4.0 International License (http://creativecommons.org/licenses/by/4.0/), which permits use, sharing, adaptation, distribution and reproduction in any medium or format, as long as you give appropriate credit to the original author(s) and the source, provide a link to the Creative Commons license and indicate if changes were made.

The images or other third party material in this chapter are included in the chapter's Creative Commons license, unless indicated otherwise in a credit line to the material. If material is not included in the chapter's Creative Commons license and your intended use is not permitted by statutory regulation or exceeds the permitted use, you will need to obtain permission directly from the copyright holder.

The Spaces of Local Climate Action

How Have Climate Emergency Declarations Helped Local Government Action to Decarbonise?

James Dyson and Calum Harvey-Scholes

Highlights Commitments within local government CEDs chart a course for faster community-level decarbonisation with participatory democracy. To move forward faster, local approaches must be equitable, coordinated and sufficiently resourced and empowered.

Keywords Climate emergency declarations · Framing · Enabling environment · Engagement · Ambition · Local government

J. Dyson (✉)
Grantham Research Institute, LSE, London, UK

Shoreham-By-Sea, Ricardo, UK

London School of Economics, London, UK

C. Harvey-Scholes
Energy Policy Group, University of Exeter, Penryn Campus, Penryn, UK

© The Author(s) 2022
C. Howarth et al. (eds.), *Addressing the Climate Crisis*,
https://doi.org/10.1007/978-3-030-79739-3_5

INTRODUCTION

The reduction of carbon emissions is urgently needed to avoid climate breakdown and local government has a critical role in delivering it (CCC, 2020a). Local governments can lead and coordinate climate action, establishing a local enabling environment for emissions reduction across the public and private sectors, and ensuring efforts are informed by meaningful engagement with people. This chapter presents analysis of the more than three hundred UK local governments who have now declared a 'climate emergency', drawing on interviews as well as the declaration texts themselves. This chapter discusses whether the emergency rhetoric and declarations have accelerated policy change, how they may affect the role of local government in decarbonisation, and what could help local governments reduce emissions faster.

The Role of Local Climate Governance in the UK

The importance of the role of local government in achieving the UK's goal of an ambitious 63% reduction in carbon emissions compared to 2019 levels by 2035 is increasingly recognised (Amundsen et al., 2018; BEIS, 2020; CCC, 2020b). However, the past decade of austerity has seen central government funding for local government halve, meaning a 27% real terms reduction in spending power (National Audit Office, April 2019). In addition, the withdrawal of the National Reporting Framework for local government in 2011, which included environmental metrics, has weakened local government's ability to address the low carbon agenda (Dixon & Wilson, 2013). Consequent reductions in environmental policy capacity in UK local governments (Eckersley & Tobin, 2019) have resulted in considerable variation in activity to implement sustainable energy systems across the country (Britton, 2018; Tingey & Webb, 2020). Finances and limitations to political authority both constrain local governments' capacity for action, and overcoming current circumstances has required innovation from local governments (Kuzemko & Britton, 2020).

The Growth of Local Climate Action

Despite these challenges, the past two years have seen a groundswell in local climate action around the UK and many local governments have shown real ambition to accelerate the UK's journey to net-zero. 2019 saw interest in and concern for climate change among the UK public reach their highest in decades (Skinner, 2019; UNDP and University of Oxford, 2021). Climate Emergency Declarations (CEDs) have now been passed by the majority of UK Tier 1 and 2 local governments (such as borough or county councils, unitary authorities and district councils), in many cases influenced by supportive civil society and resident groups. These can be used as an indicator to map the short-term local government policy response to this grassroots movement for climate action, as shown in Fig. 5.1.

Local government CEDs have contributed to normalising a new, more urgent 'frame' for climate change. The declarations redefine climate change from a technical challenge to a narrative which leverages the most recent climate science to emphasise an urgent 'climate emergency' in need of rapid redress. CEDs also emphasise the human consequences of failing to address climate change, stressing the moral responsibility of local government to act. The changing climate is already making life more challenging for people living in vulnerable regions, including small holder farmers (Harvey et al., 2018) and small island states (Monioudi et al., 2018), whilst the threat of runaway climate change is causing an intensifying cognitive burden for younger generations (Clayton, 2020). These ideas are visible in the CED texts. In making these declarations, local

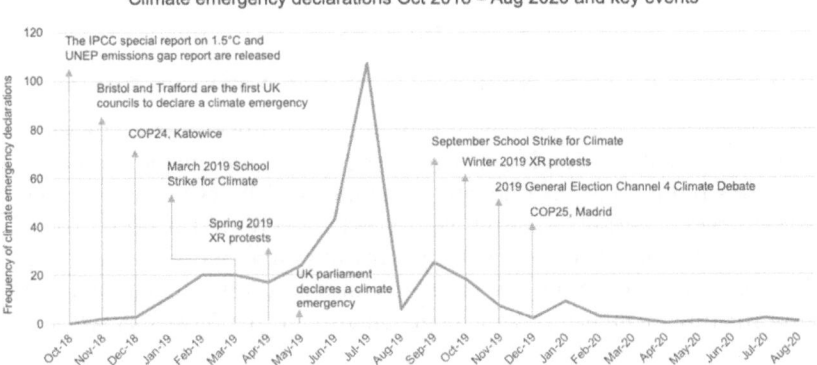

Fig. 5.1   CEDs by month, October 2018 to August 2020 with key events

governments have helped to normalise the urgency of the problem and (re)positioned themselves as centrally engaged in tackling it.

The majority of UK local government CEDs had a focus on climate mitigation and were substantiated with ambitious net-zero targets and commitments to take the steps needed to reach those targets. Furthermore, most declarations and plans made public participation, often involving young people, a core aspect of devising their routes to net zero (Grantham Institute, 2021). In the year or so that has followed the CEDs, progress on the commitments includes (as of January 2021) 225 of 404 UK local governments having produced new climate action plans to reach net-zero targets. Two-thirds of these aim for a net-zero target ahead of the government's 2050 target (a large number of these aim for 2030).

The arguments presented in the rest of this short chapter are based on the outcomes of the authors' research projects[1] which analysed the content of over 300 UK CED texts and supplemented this with a macro-analysis of every local government's climate ambitions as of 2020. In addition, 48 councillors, council officers and local residents from 26 UK local authorities across England and Scotland were interviewed between July 2019 and July 2020 on how they viewed the role to be played by these declarations. Analysis of this broad set of data provides insights into the nature of the CEDs as well as the social and policy environment in which they emerged.

An 'Enabling Environment' for Local Emissions Reduction

Charting a Course: Motivation, Direction-Setting, and Targets

We argue that the CEDs have prompted local government, to varying degrees, to establish conditions which encourage and enable local carbon emissions reductions. As one council employee observed, 'It's definitely stepped everything up .... And we needed it really [...] because it would have been business as usual'. Through gearing internal processes to deliver plans and achieve ambitious targets, utilising governmental policy and authority and facilitating public engagement and participation, whilst not

[1] Research was part of an MSc thesis, ESRC-funded PCAN work and the EPSRC-funded IGov project.

comprehensive, local authorities can be seen to be creating an 'enabling environment' (Bulkeley & Kern, 2006).

Elected officials and officers in local government expressed the feeling that the declarations empowered the councils to take action—showing people within councils that they could control their own futures and decisions on this issue. Councillors, local government officers and civil society members spoken to, believe the declarations have been effective tools to add formality and legitimacy to the need and mandate for faster action, making climate 'more of a mainstream issue that everyone has to think about'. The declarations disrupted the social fabric of local governments, raising the profile of the issue and empowering officials to act. This empowerment has enabled newly ambitious climate strategies, such as creating an action plan, ensuring appropriate and effective governance, and monitoring progress. Of course, whilst these foundational steps represent an advance on the status quo, their translation into practice and the acceleration of decarbonisation will ultimately determine success.

In addition, many CEDs directly committed local governments to institutional reforms: such as, establishing climate responsibilities in Cabinet and other committees, adding a climate impact assessment to their decision-making processes, ensuring strategic policy alignment with zero carbon and running local engagement forums. For example, Manchester City Council have introduced a 10% weight for climate change in their decision-making framework (CCC, 2020a). Looking to influence building development in the wider area, others have begun a review of their Local Plan in order to align it with their CED (e.g. Basingstoke & Deane Borough Council, Cotswolds District Council, and Stafford Borough Council). Overall, normalising and formalising the climate emergency within local government processes and activity, as well as regulating and steering others, provides a clarity of purpose and a shared trajectory for emissions reduction.

Public Engagement: Consent and Accountability

Civil society has played a role in elevating climate on the policy agenda and promoting the climate emergency frame—consider the widely-publicised Extinction Rebellion occupations, the Fridays for Future school strikes and Greta Thunberg's blunt candour in addressing those in authority. More directly, we have observed residents' groups initiating the CED process and frequently collaborating with elected officials, encouraging them to raise motions and demonstrating public support.

The CEDs commit to engage with the community, often with young people specifically; some invoke youth parliaments and at least 18 have already carried out citizens' assemblies. Deliberative engagement exercises such as citizens' assemblies and juries have emerged as a newly prominent feature in policy development—examples include Camden Borough Council, Kendal Town Council and Oxford City Council. Increasing deliberative exchange between civil society and policymakers has the potential to establish a social mandate for action and engage new social groups in local climate action (Howarth et al., 2020). The way the council reached out to citizens ahead of declaring a climate emergency suggests how CEDs involved new people and politicians in local climate action. As described by one resident:

> Jane's name was pulled out. She wasn't initially a climate activist and was more kind of from a strong trade union background. She spoke to a lot of people in her ward that had started to do work on the climate crisis.

Effective publicity of these (arcane and novel) processes is important for public buy-in as well as an opportunity for raising awareness of the climate issue more widely.

There are early, promising signs that the increased engagement between civil society and policymakers can continue, with citizens monitoring policy progress and expressing their (dis)approval. In an early example of ongoing engagement, Bury Council received sharp criticism in early 2020 for not making good on the promises made in their CED, leading the council to revise their budget. Ongoing accountability will partly depend on the extent to which local civil society organisations hold their elected officials to account.

Moving Forward Faster

Declaring a climate emergency signals a direction of travel and, in many cases, indicates the required speed. Progress requires practical policy to drive carbon emissions reduction. Pushing the metaphor a little further, in order to arrive in time, action must accelerate—the current cruise control simply will not get us to the destination in time to avoid catastrophic heating (UNEP, 2020). As we have seen, preliminary activity has prepared an environment for action within institutions and, importantly, begun engagement with businesses and residents to co-create solutions and aspirations. In order to deliver faster progress, local governments may explore a more active role in certain areas (Britton, 2018). For example, local governments have led the development of district heat networks for decades in the UK (e.g. Sheffield City Council, Islington Borough, Gateshead Council) and successful municipal local transport provision persists (e.g. Lothian Buses, Reading Buses), with the idea being considered elsewhere (e.g. Cornwall). Bristol City Council's *Bristol City Leap* project aims to leverage private sector investment to deliver a city-scale low-carbon energy transformation programme. Given that many of the technological solutions are already at commercial scale, if the CEDs are interpreted as a genuine ambition to enact transformative policy, then the challenges local government faces in achieving the demanding rate of emissions reduction required by the science are primarily social, practical and political. More specifically, on the basis of our research, we identify three requirements for accelerating emissions reduction: public buy-in to policy and the wider agenda, coordination between and across scales to deliver on commitments, and the power and investment to act. We now briefly consider each of these in turn.

The Limits to the Emergency Frame

In order to include all groups in climate discussion and decision-making, public engagement communications must be tailored and often expressly related to diverse social, financial, and other local circumstances, as well as environmental benefits.

Our research identified a concern that 'climate emergency' may not be an engaging frame for people who are from socially and economically marginalised groups. For some people, declaring an emergency for climate change seems irrelevant or even marginalises the everyday emergencies that they experience. For those experiencing daily hardship such as precarious income, chronic illness or unstable housing, being able to treat climate change as an emergency may be inconceivable. Reducing inequality and poverty must be a core ambition within climate action strategies. The challenge of decarbonisation presents many opportunities to tackle both issues, simultaneously reducing hardship and carbon emissions, including retrofitting homes and reducing car traffic (MacNaughton et al., 2018; Sharifi, 2021).

Coordinating Action

A combination of central coordination and local collaboration will be essential to meeting ambitious emissions reductions targets. Coordination across local governments is both challenging (Clar, 2019) and vital to achieving the goals set out in the CEDs (Bulkeley & Kern, 2006; CCC, 2020a). Between local governments, devolved regions and the national governments, net-zero targets vary significantly; these varying aspirations are likely to cause challenges for coordinating action between local governments as well as with national government. A balance between local autonomy and central governance will help in areas such as energy system transformation (Willis et al., 2019). Evidence of innovative solutions is emerging; for instance, the Devon Climate Response Group aims to bring actors across the county together under a common net-zero vision and coordinated action plan. Local partnerships are forming, with the aim to coordinate climate action among businesses and communities, such as the Place-Based Climate Action Network (PCAN) commissions. Research and demonstrations to enhance our understanding of effective coordination across governmental scales, energy vectors and sectors are needed.

The Available Power and Resource for Local Government

Some local governments may partly overcome resource constraints, but fulfilment of ambitious CED commitments across the board will be expedited by financial support and devolution of powers from central government. An appeal to central government to provide the power and resource required to deliver on commitments is a common feature among the CEDs. Though policy to enable local action has been promised by the government (Hansard, 2019), delivering this will require both clear strategic direction-setting from Whitehall at the same time as providing autonomy to implement appropriate local solutions through powers and finance. Nonetheless, innovative local governments are finding ways to finance action in the absence of directly supportive central government policy: for example, Warrington and West Berkshire councils became the first councils to successfully lever low-cost private finance for renewable generation projects using 'green bonds' (WBC, 2021). Warwick Council, meanwhile, has announced a referendum asking residents whether they would accept increased council tax to fund climate action.

REFERENCES

Amundsen, H., et al. (2018). Local governments as drivers for societal transformation: Towards the 1.5°C ambition. *Current Opinion in Environmental Sustainability, 31*, 23–29. https://doi.org/10.1016/j.cosust.2017.12.004

BEIS. (2020). *Energy white paper: Powering our net zero future.*

Britton, J. (2018). Localising energy: Heat networks and municipal governance. In *Handbook of international political economy of energy and natural resources.* Edward Elgar Publishing.

Bulkeley, H., & Kern, K. (2006). Local government and the governing of climate change in Germany and the UK. *Urban Studies, 43*(12), 2237–2259.

CCC. (2020a). *Local authorities and the Sixth Carbon Budget.* https://doi.org/10.1016/s0033-3506(44)80323-7

CCC. (2020b). *The Sixth Carbon Budget—The UK's path to net zero.* Available at: https://www.theccc.org.uk/publication/sixth-carbon-budget/

Clar, C. (2019). Coordinating climate change adaptation across levels of government: The gap between theory and practice of integrated adaptation strategy processes. *Journal of Environmental Planning and Management, 62*(12), 2166–2185.

Clayton, S. (2020). Climate anxiety: Psychological responses to climate change. *Journal of Anxiety Disorders, 74*, 102263.

Dixon, T., & Wilson, E. (2013). Cities' low-carbon plans in an "age of austerity": An analysis of UK local authority actions, attitudes and responses. *Carbon Management, 4*(6), 663–680. https://doi.org/10.4155/cmt.13.58

Eckersley, P., & Tobin, P. (2019). The impact of austerity on policy capacity in local government. *Policy & Politics, 47*(3), 455–472.

Hansard. (2019). *Oral answers to questions 20th March, Hansard, volume 656.* Available at: https://hansard.parliament.uk/Commons/2019-03-20/deb ates/9966CB46-DA75-40C2-BB07-8294CE9F4348/OralAnswersToQuest ions#contribution-84F3A0B0-3106-4DE7-9473-F87BDB8B2E91. Accessed 8 January 2021.

Harvey, C. A., et al. (2018). Climate change impacts and adaptation among smallholder farmers in Central America. *Agriculture & Food Security, 7*(1), 57. https://doi.org/10.1186/s40066-018-0209-x

Howarth, C., et al. (2020). Building a social mandate for climate action: Lessons from COVID-19. *Environmental and Resource Economics, 76*(4), 1107–1115. https://doi.org/10.1007/s10640-020-00446-9

Kuzemko, C., & Britton, J. (2020, November 2019). Policy, politics and materiality across scales: A framework for understanding local government sustainable energy capacity applied in England. *Energy Research and Social Science, 62*, 101367. https://doi.org/10.1016/j.erss.2019.101367

MacNaughton, P., et al. (2018). Energy savings, emission reductions, and health co-benefits of the green building movement. *Journal of Exposure Science &*

Environmental Epidemiology, 28(4), 307–318. https://doi.org/10.1038/s41 370-017-0014-9

Monioudi, I. N, et al. (2018). Climate change impacts on critical international transportation assets of Caribbean Small Island Developing States (SIDS): The case of Jamaica and Saint Lucia. *Regional Environmental Change, 18*(8), 2211–2225. https://doi.org/10.1007/s10113-018-1360-4

Sharifi, A. (2021). Co-benefits and synergies between urban climate change mitigation and adaptation measures: A literature review. *Science of The Total Environment, 750,* 141642. https://doi.org/10.1016/j.scitotenv.2020. 141642

Skinner, G. (2019). *Concern about climate change reaches record levels with half now 'very concerned', Ipsos Mori.* Available at: https://www.ipsos.com/ipsos-mori/en-uk/concern-about-climate-change-reaches-record-levels-half-now-very-concerned. Accessed: 11 January 2021.

Tingey, M., & Webb, J. (2020). Governance institutions and prospects for local energy innovation: Laggards and leaders among UK local authorities. *Energy Policy, 138,* 111211. https://doi.org/10.1016/j.enpol.2019.111211

UNDP and University of Oxford. (2021). *Peoples' Climate Vote.* Available at: https://www.undp.org/content/undp/en/home/librarypage/climate-and-disaster-resilience-/The-Peoples-Climate-Vote-Results.html

UNEP. (2020). *Emissions Gap Report 2020.* Nairobi.

WBC. (2021). *West Berkshire Community Municipal Investment* (CMI). West Berkshire Council. Retreived from: https://info.westberks.gov.uk/wbcmi.

Willis, R., et al. (2019). *Enabling the transformation of the energy system, IGov Project.* Available at: http://projects.exeter.ac.uk/igov/wp-content/upl oads/2019/04/Enabling-the-transformation-of-the-energy-system-DRAFT-FOR-COMMENTS.pdf

Open Access This chapter is licensed under the terms of the Creative Commons Attribution 4.0 International License (http://creativecommons.org/licenses/by/4.0/), which permits use, sharing, adaptation, distribution and reproduction in any medium or format, as long as you give appropriate credit to the original author(s) and the source, provide a link to the Creative Commons license and indicate if changes were made.

The images or other third party material in this chapter are included in the chapter's Creative Commons license, unless indicated otherwise in a credit line to the material. If material is not included in the chapter's Creative Commons license and your intended use is not permitted by statutory regulation or exceeds the permitted use, you will need to obtain permission directly from the copyright holder.

CHAPTER 6

Developing a Carbon Baseline to Support Multi-Stakeholder, Multi-Level Climate Governance at County Level

Erica Russell and Ian Christie

Highlights Orchestration requires political commitment and engagement on the basis of evidence, knowledge and progress-checking. Local actors face challenges in compiling carbon baselines that offer useful production and consumption emissions.

Keywords County · Baseline · Carbon emissions

THE IMPORTANCE OF LOCAL BASELINE DATA

There is widespread acceptance that top-down approaches to climate change, identified with the Kyoto Protocol (Jordan et al., 2018, p.4), are no longer sufficient to drive climate action. International bodies,

E. Russell (✉) · I. Christie
Centre for Environment and Sustainability (CES), University of Surrey, Guildford, UK
e-mail: erica.russell@surrey.ac.uk

© The Author(s) 2022
C. Howarth et al. (eds.), *Addressing the Climate Crisis*,
https://doi.org/10.1007/978-3-030-79739-3_6

such as the UN, are seen as providing global direction, but with the Paris Agreement came an acceptance that implementation based on 'real world' experimentation required greater action by state, sub-state and non-state organisations (Oberthür, 2016). This shift in thinking has seen increased debate about effective governance forms: those focused on mutually interdependent national and sub-national actors; the multi-level governance approach first identified by Hooghe and Marks (1996); and the related concept of polycentricity, which focuses on local leadership through self-coordinating groups, often as part of wider networks (Ostrom, 2014; Backstrand et al., 2018). Increasingly, a need for both approaches has been cited, as capacity, resources and reach need to be shared (Newell et al., 2012). It is within a context of frequently 'unco-ordinated' sub-national action (Bansard et al., 2017) that this chapter considers a strategically significant issue for climate policymaking and for 'orchestration' of climate governance (Backstrand et al., 2018)—namely, the difficulties of creating an effective *emissions baseline* suitable for local actors to use as a basis for climate mitigation planning and implementation. Specifically, we consider the *county* level of local action in the UK, focusing on Surrey, a county in England.

Establishing a Carbon Baseline for Surrey

Carbon footprinting and baselining exercises have been completed for many *cities*, including several in the UK, but little research has been undertaken at *larger sub-national scales*. This case study offers insights from the carbon baseline work initiated by the Surrey Climate Commission and undertaken by the Centre for Environment and Sustainability at the University of Surrey. (This exercise complements recent carbon footprinting work carried out for Surrey County Council.) The Surrey experience sheds light on issues arising in efforts to provide crucial climate-related information for a territory that includes large urban populations, extensive suburban areas and a substantial rural population and area. Working at the scale of a county creates both complexity and opportunities. Surrey is adjacent to London, with a population of approximately 1.2 million who live in its 26 towns, 175 villages and hamlets. The county

I. Christie
e-mail: i.christie@surrey.ac.uk

comprises large areas of downland and sandy heath, and is highly wooded (22% of area), and farming tends to be extensive.

Surrey is administered through multiple tiers: a county council, 11 borough and district councils, and more than 80 parish and town councils and is part of two wider sub-regional Local Enterprise Partnerships. Surrey also has strong advocacy groups, with over 50 organisations involved in environmental or climate activities (Street, 2020). Whilst this degree of institutional richness may be a local strength, supporting both multi-level and polycentric approaches to climate governance, risks arise. Without a clear vision for coordination and long-term planning, such a plethora of actors can result in confused responsibility and reduced impact in environmental and climate policy (Newell, Pattberg & Schroeder, 2012). This complexity poses challenges in carbon baseline studies distinct from those arising for cities.

Drawing together this diversity of actors and county attributes, the Surrey Climate Commission provided a leadership role, acting as both the initiator and an independent actor (Homsy & Warner, 2015) in requiring a baseline study. Accepting that limited information results in poorly targeted climate action plans (Boehnke et al., 2019; Lehtonen & Kern, 2009), a local and relevant emissions baseline was seen as critical for highlighting carbon 'hotspots'. Failure to overcome deficiencies in localised data and action planning was also identified as restricting the development of best practice (Boehnke et al., 2019). As a result of these constraints, there is little evidence that increased *capacity* for local climate action has resulted in actual *reductions* in local carbon emissions (Hoppe et al., 2016), a situation the Surrey Climate Commission wanted to address.

With political and financial limitations in mind, a primary aim of the baseline research was to use readily accessible publicly available data that would allow for ease of ongoing monitoring at little additional cost or expertise. Where possible, the research utilised sub-national emissions datasets (BEIS, 2020b) to provide quality assurance and to align with the local authority reporting frameworks. Working at the county scale, the baseline had to consider land use, with its potential for carbon capture, high levels of variation in district profiles, both physical and population based, and in the case of Surrey, the impact of London commuting and wealth transfer. Key to the engagement of local actors was the provision of a baseline carbon footprint that offers this nuanced understanding of place, local issues and interest group alignment. Whilst local climate action by public bodies has focused primarily on territorial emissions, the Surrey Climate Commission's members made it clear, through a process of consultation, that the baseline work must additionally incorporate and highlight the impacts of *consumption* as well as of local emissions from production.

ISSUES ENCOUNTERED IN CREATING A USEFUL COUNTY-SCALE BASELINE

The Surrey Climate Commission baseline research project has identified several issues in creating county-scale baselines that we expect would face similar county or sub-regional level work. Most importantly, national datasets, even those available at a sub-national level, are based on international emissions reporting commitments and national government policy needs. It is clear that multiple reporting formats have created discrete UK carbon datasets. Some of these are spatially separated, and others use different methodologies, data and extensive modelling to provide insight into specific sectors or issues. Even direct energy use data lack granularity, with BEIS acknowledging that an annual spend threshold may mean a misallocation of up to 2 million small businesses as domestic users (NAEI, 2020). Information is also held in different measurement units and carbon formats. Such variation in methodologies makes direct comparison difficult.

UK Sub-National Consumption statistics (BEIS, 2019b) provide emissions data for four fuel categories: electricity, gas, other heating fuels and transport fuel, allocated across three territorial categories: domestic, industrial/commercial and transport. Additionally, they provide data on land use, for both carbon emissions and sequestration. All emissions are supplied in units of CO_2 and are available at both county and district levels. This information provides a useful guide to county-based carbon hotspots. However, the UK Carbon Footprint (DEFRA, 2019), based on models using value flows, is currently only available at a *national* level.

ADDRESSING BASELINE LIMITATIONS

To overcome the disconnect between this top-down data availability and the types of information needed to support practical, local action, the researchers worked with the Surrey Climate Commission to identify key areas to expand as part of the Surrey baseline:

1. Enhanced spatial and use detail of people's homes and Surrey travel;
2. Identify the local impact of business and the public estate;
3. Increase understanding of the land and its role in carbon sequestration;
4. Estimate the size of the county's carbon footprint.

In doing this, an important principle was established: namely, that with increased granularity came a coarsening of the data, but that this trade-off was acceptable if it provided richer insight, supported proportionality of response, identified gaps and made visible unseen issues. The following sections provide examples of this work.

Creating Richer Insight

The highest territorial emissions in Surrey are associated with transport and travel (50.2%). Sub-regional data (BEIS, 2020b) confirm that traffic on Surrey's A-roads and motorways generate the greatest emissions, but provide no detail on what types of vehicles are creating the emissions—or why the vehicles are being driven. The most detailed information on work patterns and commuting at an individual level is the Census dataset. Using this, it was possible to understand local work and commuting patterns by distance and transport type, albeit with the caveat that this information is now dated. The data are even less reliable as a guide to the future as a result of COVID-19, which has expedited changes in shopping patterns and an increase in working from home, shifts which are unlikely to be completely reversed after the virus effect is overcome. Car usage was the primary generator of emissions on all types of roads, creating between 52 and 70% of Surrey districts' transport emissions. Combining 'reason to travel' national survey data (DfT, 2018) with calculated Surrey car emissions enabled a crude allocation of Surrey resident travel. This suggested that whilst home working could reduce commuting-related emissions, up to 33% of car-based emissions were generated in visiting friends and family. Here, reduction may require low carbon travel alternatives.

As noted earlier, information on domestic electricity and gas emissions is available at a tier 2 level but additional work using domestic Energy Performance Certificate (EPC) data (MHCLG, 2019) provided a more nuanced guide to the types of homes in each district and average emissions. This analysis indicated a strong correlation between house size, affluence and higher emissions usage. Whilst many councils have focused on the social co-benefits achieved by supporting those in fuel poverty, this work identifies that there is an ongoing need to promote behavioural change among those citizens most able to afford carbon reduction.

Offering Perspective on the Scale of Emissions

Many public bodies have taken on a leadership role in decarbonisation in their own estate, promoting energy demand reduction, testing new technologies at scale or undertaking large exemplar renewable projects. It is important that local actors have a realistic perspective on the direct impact such activity can have at a *county* level. This is difficult, as BEIS sub-national data do not differentiate public sector emissions from those of industry and commerce. Our attempts to estimate county emissions from the public sector, using locally available data for Surrey, were only partially successful. Whilst the County Council, the University of Surrey and NHS primary healthcare sites could provide annual emissions data, those for district councils were incomplete; and information on emissions from secondary healthcare sites was extremely limited. We concluded that the public sector accounted for around 2% of the county's total emissions, although it is likely that this is an underestimation. It is, however, in line with published UK national public sector emission estimates (BEIS, 2020a).

For local organisations wanting to drive change within industry and commerce (19.3% of Surrey CO_2 emissions), where and why these emissions occur remains a difficult question to answer. Whilst ONS sub-national data include energy and travel emissions for the agricultural sector, ONS offers no granularity for other sectors. To overcome this, our baseline work drew together national business emissions for both CO_2 and CO_2e by sector (ONS, 2019a), national business numbers by sector (BEIS, 2019a) and numbers of businesses by sector in Surrey (ONS, 2019b). Carbon dioxide emission data suggest that manufacturing businesses create the highest sectoral emissions (45%) across all but one of the county's districts, whilst the logistics sector (22.6%) is significant in one of the districts. The baseline study also reviewed industry CO_2e emissions: here it seems that there is considerable under-reporting of emissions within agriculture, forestry and fishing, and more substantially in the manufacturing sector, where more than 80% of emissions identified are derived from gases other than CO_2. Other sectors are less affected, appearing to emit around 90% of their carbon as CO_2.

Identifying Gaps

In 2018, ONS data confirmed that Surrey land acts as a carbon sink. Expanded information available at a local authority level now provides positive and negative emissions from four types of land use (BEIS, 2020b). This offers increased granularity of data on carbon sequestration due to local land use change, indicating the benefit of increasing woodland, peat wetlands and grassland. However, the dataset only enables high level monitoring of land use change, which limits the use of data in informing strategy and driving action. This gap is being investigated further.

Making Hidden Impacts Visible

To understand the 'hidden' carbon impact of products and services bought by those living and working in Surrey, the research attempted to allocate national footprint data. Simple pro-rata allocation by population size did not allow for the impact of affluence, an issue highlighted in the Surrey homes data, which would result in an underestimation of consumption emissions. Therefore, we adapted and updated the work of Minx et al. (2013), who combined both MRIO data with information on a variety of metrics linked to affluence: this approach suggests a carbon footprint of 16,898 ktCO$_2$. Whilst a relatively crude allocation, this would certainly suggest that Surrey's overall carbon footprint, combining the production and consumption perspective, is at least *twice* the size of the territorial emissions.

CONCLUSION

With an acceptance that climate change action is a responsibility of all, we argue that polycentric approaches need to be underpinned by knowledge at all levels. The challenges of place-based climate action at local levels in the UK and beyond are multifaceted, and effective action depends on a good base of knowledge to help decision-makers navigate the complexity. Whilst there is much to welcome in ground-up action, we suggest that a level of orchestration is required. National datasets need to be improved as indicated in our earlier discussion. But a crucial additional task for central government with actors at sub-national scale is to ensure that national data are complemented by adequate resources to enable local authorities and their partners to establish and update datasets on sectoral emissions at city, county and district/borough level. We suggest too that work needs to be done on development and take-up of a standard set of carbon mapping tools and metrics at local scales, to enable comparisons, collaborations and information exchange between actors in climate governance at local and regional levels. Finally, urgent work is needed on measuring progress in reducing emissions from consumption. Given the extent of diversity and inequality in local economic and social conditions, we suggest there is great value in locating that work primarily at local levels. We recommend that central government equip a variety of local authorities to act as centres of excellence in mapping and measuring progress in reductions in lifestyle-based emissions (these areas could well be drawn from those that have set up PCAN Climate Commissions).

References

Bäckstrand, K., Zelli, F., & Schleifer, P. (2018). The legitimacy and accountability in polycentric climate governance. In A. Jordan, D. Huitema, H. van Asselt, & J. Forster (Eds.), *Governing climate change: Polycentricity in action* (pp. 338–356). Cambridge University Press. https://doi.org/10.1017/978110828464 6.020

Bansard, J. S., Pattberg, P. H., & Widerberg, O. (2017). Cities to the rescue? Assessing the performance of transnational municipal networks in global climate governance. *International Environmental Agreements: Politics, Law and Economics, 17*(2), pp. 229–246.

BEIS. (2019a). *Business population estimates 2019*. Available at: https://www.gov.uk/government/statistics/business-population-estimates-2019. Accessed: 29 January 2020.

BEIS. (2019b). *Sub-national consumption statistics methodology and guidance booklet*. London. Available at: https://www.gov.uk/government/publicati ons/regional-energy-data-guidance-note

BEIS. (2020a). *2018 UK greenhouse gas emissions, final figures*. London. Available at: https://assets.publishing.service.gov.uk/government/uploads/sys tem/uploads/attachment_data/file/862887/2018_Final_greenhouse_gas_ emissions_statistical_release.pdf

BEIS. (2020b). *UK local authority and regional carbon dioxide emissions national statistics: 2005 to 2018*. London: BEIS. Available at: https://www.gov.uk/ government/statistics/uk-local-authority-and-regional-carbon-dioxide-emissi ons-national-statistics-2005-to-2018

Boehnke, R. F., et al. (2019). Good practices in local climate mitigation action by small and medium-sized cities; exploring meaning, implementation and linkage to actual lowering of carbon emissions in thirteen municipalities in The Netherlands. *Journal of Cleaner Production, 207,* 630–644.

DEFRA. (2019). *UK's carbon footprint 1997–2016.* London. Available at: file:///C:/Users/User/Documents/UK Carbon Footprint 1997–2016.pdf.

DfT. (2018). *National travel survey: England 2018.* London. Available at: https://assets.publishing.service.gov.uk/government/uploads/system/uploads/attachment_data/file/823068/national-travel-survey-2018.pdf. Accessed 9 June 2020.

Homsy, G. C., & Warner, M. E. (2015). Cities and sustainability: polycentric action and multilevel governance. *Urban Affairs Review, 51*(1), 46–73.

Hooghe, L., & Marks, G. (1996). "Europe with the regions": Channels of regional representation in the European Union. *Publius: The Journal of Federalism, 26*(1), 73–92.

Hoppe, T., Van der Vegt, A., Stegmaier, P. (2016). Presenting a framework to analyze local climate policy and action in small and medium-sized cities. *Sustainability, 8*(9), 847.

Jordan, A., et al. (2018). *Governing climate change: Polycentricity in action?* Cambridge University Press.

Lehtonen, M., & Kern, F. (2009, Springer). Deliberative socio-technical transitions. In *Energy for the future* (pp. 103–122). Palgrave Macmillan.

MHCLG. (2019). *Table EB7: Domestic Energy Performance Certificates for existing dwellings by type of property, average energy use, carbon dioxide emissions and fuel costs, Statistical data set Live tables on Energy Performance of Buildings Certificates.* Available at: https://www.gov.uk/government/statistical-data-sets/live-tables-on-energy-performance-of-buildings-certificates. Accessed 18 February 2020.

Minx, J., et al. (2013). Carbon footprints of cities and other human settlements in the UK. *Environmental Research Letters, 8*(035039), 1–10.

NAEI. (2020). *Local and regional carbon dioxide emissions estimates for 2005–2018 for the UK.* London. Available at: https://www.gov.uk/government/statistics/uk-local-authority-and-regional-carbon-dioxide-emissions-national-statistics-2005-to-2017

Newell, P., Pattberg, P., & Schroeder, H. (2012). Multiactor governance and the environment. *Annual Review of Environment and Resources, 37.*

Oberthür, S. (2016). Reflections on global climate politics post Paris: Power, interests and polycentricity. *The International Spectator, 51*(4), 80–94.

ONS. (2019a). *Atmospheric emissions: greenhouse gases by industry and gas: Total greenhouse gas emissions by industry section, 1990 to 2017 and (provisional) 2018.* Available at: https://www.ons.gov.uk/economy/environmentalaccounts/datasets/ukenvironmentalaccountsatmosphericemissionsgreenhousegasemissionsbyeconomicsectorandgasunitedkingdom. Accessed: 29 January 2020.

ONS. (2019b). *Number of businesses in Surrey*. Available at: https://www.surreyi. gov.uk/dataset/24jw6/number-of-businesses-in-surrey. Accessed: 29 January 2020.

Ostrom, E. (2014). A polycentric approach for coping with climate change. *Annals of Economics and Finance, 15*(1), 97–134.

Street, P. (2020). *Vision for sufficient action in Surrey*. Guildford.

Open Access This chapter is licensed under the terms of the Creative Commons Attribution 4.0 International License (http://creativecommons.org/licenses/ by/4.0/), which permits use, sharing, adaptation, distribution and reproduction in any medium or format, as long as you give appropriate credit to the original author(s) and the source, provide a link to the Creative Commons license and indicate if changes were made.

The images or other third party material in this chapter are included in the chapter's Creative Commons license, unless indicated otherwise in a credit line to the material. If material is not included in the chapter's Creative Commons license and your intended use is not permitted by statutory regulation or exceeds the permitted use, you will need to obtain permission directly from the copyright holder.

CHAPTER 7

Power in Practice: Reflecting on the First year of the Edinburgh Climate Commission

Rosanna Harvey-Crawford and Alice Creasy

Highlights Insiders' perspectives on the first year of the Edinburgh Climate Commission. Innovative governance, or the reproduction of existing power structures?

Keywords Local governance · Climate change · Cities · Net-zero · Power · Ethnography

INTRODUCTION

Surfing a wave of place-based urban governance, the creation of Climate Commissions through the Place-based Climate Action Network (PCAN) reflects an increasing focus on cities as 'strategic sites' for experimental

R. Harvey-Crawford (✉) · A. Creasy
School of GeoSciences, University of Edinburgh, Edinburgh, UK
e-mail: r.e.harvey-crawford@ed.ac.uk

A. Creasy
e-mail: alice.creasy@ed.ac.uk

© The Author(s) 2022
C. Howarth et al. (eds.), *Addressing the Climate Crisis*,
https://doi.org/10.1007/978-3-030-79739-3_7

climate governance (Broto, 2019). The creation of these Commissions under the broad umbrella of 'PCAN', however, means that these new groups are caught up in a complex, multi-scalar and fast-evolving landscape which connects the messy and lived experiences at 'the local' to a much broader network of actors and institutions across the country, and across the world. It is this a tangled landscape that those charged with 'setting up' these place-based Commissions must try, not only to interoperate, but to work within.

This chapter reflects on the first year of the *Edinburgh* Climate Commission: from the 'setting up' process in Autumn/Winter 2019 to November 2020 when the Commissioners were looking forward to the next phase of their 2020 workplan. This period marks a unique part of the Commission's history as stakeholders grapple with a dynamic landscape, attempting not only to define the role of the Commission but conceptualise and represent the 'place' in which it exists. In grappling with these questions, there exists an opportunity to develop a mode of climate governance which combines the input of 'organic intellectuals', whose knowledge is grounded in everyday experiences and working-class life, alongside 'traditional intellectuals', whose knowledge is grounded in formal expertise (Gramsci, 1971). This mixing of formal and informal knowledges and experiences sparks more inclusive climate actions through an articulation of knowledge that is place-based and culturally inclusive (Rice et al., 2015).

Written from the perspective of two junior, female members of the team charged with setting up the Commission, this chapter will be a personal, reflexive account of our experiences on the project as passionate environmentalists eager to become involved with local climate action in practice. Through two themes, 'The Carbon City' and 'Project Power Relations', we reflect on how power dynamics, from the personal to the international scale, have influenced the representation of place through the Edinburgh Climate Commission. Importantly, these themes articulate barriers to building a climate 'praxis' (Rice et al., 2015) and demonstrate how decisions made in the setting up phase are vitally important for shaping a Commission's future.

BACKGROUND & METHODOLOGY

The Edinburgh Commission was formally established in March 2020 at the beginning of the COVID-19 pandemic, but we (the authors of this chapter) had been part of the team working behind the scenes on the project since August 2019. This team was made up of small group of Council and University of Edinburgh staff including three core members who occupied senior roles at their respective organisations. As a recent MSc graduate and a part-time MSc student halfway through her degree, we came to the project as junior members of the team. We are both Edinburgh residents and passionate environmentalists and were excited about working on an innovative project related to the city's recent Climate Emergency declaration and pledge to be Net Zero by 2030. Rosanna was tasked with the more day-to-day 'doing' in the project—administrative tasks and attending meetings—whilst Alice was providing research support to inform the shape and structure of the Commission based on evidence from local climate action initiatives taking place elsewhere.

This chapter's contribution is based on ethnographic diaries kept by both authors since August 2019. Ethnographic research methods can be defined simply as the process of the researcher immersing themselves in the research setting: observing events, participating in conversations, examining speech for underlying assumptions and recording observations in a field diary (de Volo & Schatz, 2004). Although first used in research projects on 'traditional communities', the benefits of using ethnography to 'study up' have since been noted (Wolf, 2018). Project, or participant, ethnography has been used by researchers embedded in projects or institutions to better understand how policy is produced and how projects are implemented. It is emphasised that 'ethnographies must engage with the concept of 'power', paying attention to whose voices, interests and ideas come to dominate within projects at different times and why' (Evans & Lambert, 2008; Lewis et al., 2003). With this in mind, this chapter will reflect on the power dynamics of establishing the Edinburgh Climate Commission and on the dominant voices, ideas and interests that have persisted both before and after the Commission's launch.

THE CARBON CITY

The concept of a Climate Commission is a relatively new one and, until the expansion of the PCAN project in 2019, a practice that was firmly rooted in the geography of Leeds. As PCAN grew to encompass two more cities in late 2019, coordinators looked around for mobile pieces of policy, research and best practice that could be applied to the development of Commissions in different places. In Edinburgh, faced with not only a blank slate but a self-conscious pressure to get the Commission 'set up' and to work with the City Council, stakeholders leaned heavily on a piece of work called the Carbon Roadmap to guide the practical development of the Commission and its approach to climate change in the city. This piece of work would become integral to the development and focus of the Commission.

The Carbon Roadmap presents a carbon accounting methodology which allows researchers to develop emissions profiles for the city, broken down by sector and based on carbon budgets derived from the IPCC's global budget divided equally across the world's population. The Edinburgh iteration was aligned explicitly to the Council's target of becoming Net Zero by 2030. The methodology is based on the national Stern Review (Stern, 2006) and was led by researchers in Leeds.

The first time we saw the Carbon Roadmap, we struggled to understand it. It used a methodology we were unfamiliar with and presented a way of approaching climate action that we had not come across during our social science degrees. It was intimidating to see climate change in Edinburgh (a topic we thought we had a good grasp of) reduced to a set of numbers and sectors. While we could understand the high-level intentions of the Carbon Roadmap—technical interventions for Edinburgh to reach Net Zero—the numbers were completely inaccessible to us. The universal acceptance of this approach and its dominance across the climate change narratives being formed through the Commission left us feeling intimidated and, frankly, insecure. If this was how climate action was 'done' in the 'real world', what had we been taught at university? Did this mean what we had learnt was useless? From our perspective at the time, embedded within this project, the power that this piece of work had over stakeholders in the city and its unquestioning acceptance by those leading the Commission's development led us to assume that this was the best way to approach climate action.

As the project evolved, the Carbon Roadmap was further cemented in its importance as it formed the basis for the Council's Policy and Sustainability Committee's acceptance of the Commission in November 2019. This demonstrated the power of this carbon accounting methodology within the city, and as a result, it became the foundation on which the Climate Commission, and Edinburgh's approach to climate action, was built. From the beginning, Net Zero and the Commission have been inseparable as concepts, with one giving purpose to the other. This focus on emissions not only shaped the Commission's view of climate change as a quantitative 'issue' to be tackled by technical, measurable carbon reduction efforts but, given the 'area-based' approach to emissions reduction being touted here, served to frame the city as a bounded geography, raising questions as to what and whom would come to 'count' as *Edinburgh*. As Rice et al. (2015) observe, carbon accounting methodologies speak to a certain way of framing the climate challenge, one that privileges scientific and technical knowledge to produce exclusionary politics.

Choosing Commissioners

In November 2020, with the go-ahead from Councillors, our small team was able to start considering the task of recruiting Commission members. The goal of 'hitting the carbon numbers' (Hulme, 2019) propagated by the city's 2030 target and the Carbon Roadmap set the scene for the type of technocratic, expert knowledge that would be chosen to represent the city on the Commission. With carbon seen as a key indicator of success, the breakdown of city stakeholders by sector (private, public, third) was regarded as a 'manageable' way of making sense of the city and approaching the 'challenge' of climate change. Within these early conversations, the ability to leverage private sector capital, set out as a key mechanism for change in the Carbon Roadmap, became an important focus in drawing up the Commissioner shortlist and guiding the eventual choice of a private sector commission chair.

Instead of an application process for Commissioners, people were handpicked to fill the small number of Commission roles. The language around the selection of Commissioners was constrained by sectors, for example the need to find 'a finance representative' or 'a third sector representative'. It was at this point we began to reflect considerably on the direction this endeavour was taking and why such an approach was deemed valid. During our Masters, we had learnt about the fundamental role of social justice in addressing climate change. However, as the Commission came together, it seemed to represent one version of Edinburgh: affluent, middle class, professional. This approach highlighted how different our vision for the Commission and ideas around representation were from others working on the project. To us, representation meant the inclusion of different and diverse voices in the city, however, this perspective failed to chime with the pursuit of climate 'expertise', deemed necessary to tackle Edinburgh's emissions profile.

CO_2 has become the politically mainstream way of 'knowing' climate change. Swyngedouw (2010, pp. 219–220) has called it the 'thing' around which our 'environmental dreams, aspirations, contestations as well as policies crystallise'. Rice et al., (2015, p. 255), meanwhile, have also observed how scientific knowledge has come to dominate political discourse, noting how 'the corresponding community of technical experts that is called into importance provides a narrow pathway of understanding and action that is not sufficient to produce change because of its exclusionary politics'. In Edinburgh, the Carbon Roadmap functioned as a way of upholding a certain way of *knowing* climate change and climate action, predicated on mitigation activities grounded in leveraging finance. Through the selection of Commissioners, guided by the Carbon Roadmap, the benefits of including alternative and diverse voices were overlooked, thereby limiting the possibility of building an inclusive route to impact. Our experiences in Edinburgh thus illustrate just how powerful the narratives described by Swyngedouw and Rice (above) are in practice. Not only does carbon drive environmental policy-making (Kenis & Lievens, 2017), but it is also capable of dominating efforts to create new *institutions* for climate action.

Project Power Relations

Creating a Workplan

Due to growing pressure to act and announce the Commission's arrival in the city (further accelerated by the onset of the COVID-19 pandemic), the development and enactment of the workplan became a very controlled, closely managed process. Rather than developing the workplan with Commissioners, a draft was prepared by the Chair and secretariat and presented to them for comments. At the time, it felt as though people's lives were being turned upside down by the pandemic and that by presenting pre-prepared documents to the Commission, it would ease the burden on Commissioners, all of whom are volunteers. Beyond this, however, in these early days, there was also a sense that the secretariat (staff at the ECCI and Council) needed to maintain control of the potentially messy and unpredictable Commission in order to deliver results at pace. Consequently, the production of the workplan became overtly administrative.

As the sole administrator in the secretariat, Rosanna embodied the bureaucratic depoliticisation of what otherwise could have been an exciting political co-production exercise. Over the course of a few weeks, the Commission became a sterile, bureaucratic job that was a far cry from the dynamic, inclusive and innovative forms of climate governance we had learnt about in our Masters degrees. Most communication with the Commissioners was conducted in a flurry of emails and attached documents. It was clear that the Net Zero target and the Carbon Roadmap were seen as sufficient to guide the work of the Commission's first year, thereby preventing a Commission-wide conversation about climate action in Edinburgh where alternative views could come to the fore. Furthermore, the urgency of this target was driving a preoccupation with speed and impact that justified decisions being taken by a few key stakeholders as opposed to the wider group. Thus, the creation of the Commission (articulated through its Terms of Reference and workplan) became an increasingly closed process. Not only had the workplan served to exclude certain voices from the process, but it had also trapped us, as young, politically engaged and passionate climate activists, in administrative roles that reduced climate governance to a series of bureaucratic tasks.

Proposing a Working Group

Led by the goal of Net Zero and the Carbon Roadmap, the workplan for the Commission's first year was focused predominantly on engaging the private sector in Edinburgh, 'scaling up' action and producing a Green Economic Recovery guide for the city. Six months later, it was noticeable that some Commissioners were intimately involved with delivering the workplan, whilst there was a remaining opportunity to engage others with new ideas that more closely spoke to their skillsets. In response, we proposed a working group that could focus on public engagement and adaptation, two issues that up until then had received little attention. It felt good to take ownership and initiative in order to potentially steer the project towards climate action we felt was meaningful. Not only would this proposed working group engage Commissioners whose knowledge lay outside the private sector and engage organisations and individuals not represented on the Commission, but it would help us as employees on the project to feel empowered, useful and give us the chance to learn new skills.

Although we had buy-in from several Commissioners and from colleagues at the university, the challenge came when trying to integrate this idea with the vision of Edinburgh City Council. A key issue since the project's inception was the question of resourcing. For the Commission to deliver a demanding programme of work (outlined in the workplan), it was necessary to have more staff than the PCAN project could fund. As a result, most secretariat duties had been transferred to the Council, which would also ensure continuity once the PCAN project ended. However, resource pressure at the Council—the legacy of years of budget cuts (Ford, 2019; Centre for Cities 2019) and the COVID-19 crisis—meant that any unpredictability or deviation from the established workplan was seen as an unnecessary risk or distraction.

Because the workplan was designed and guided by a select group of individuals at an early stage, as the project evolved power remained concentrated within that same group. We experienced this power first-hand during discussions about our proposed working group, which ultimately ended up being rejected. Not only had we seen this group as an opportunity to enhance the work of the Commission in a way which would benefit the city but also a mechanism through which to empower individual people involved in the project (including ourselves). This process left us feeling disempowered and highlighted the difficulty of challenging established narratives and knowledges on how to 'do' climate action. Importantly, this experience demonstrates how difficult it is for discussions around justice, inclusivity and place-based identities to penetrate conversations on sector-based CO_2 reduction, even at the early stage of idea formulation.

Conclusion

Sustainable cities have been touted as global climate solutions, leading the way as national and international leaders fail to agree and implement effective climate policy (Angelo & Wachsmuth, 2020). With the pressure on urban centres mounting and after decades of climate inaction, a desire to act is both understandable and commendable. However, in an increasingly busy and fractured governance landscape, new climate-focused organisations must grapple with the tension between urgent action, and the ethical pitfalls of moving too fast. With this in mind, we have had a fascinating opportunity to observe and experience what this innovative leadership looks like 'on the ground' in Edinburgh.

Over the course of a year, we have seen the way in which dominant and normative narratives of climate action as Net Zero became embedded in the Commission's identity before any Commissioners had even been selected. Reflecting the global push for cities to reach Net Zero, the Carbon Roadmap laid the foundation for 'traditional intellectual' knowledge to guide the Climate Commission at the expense of building a new, grassroots and organic institution, attuned to the unique historical and geographical landscape of Edinburgh—one which could embrace the multiplicity of ways in which people can 'know' climate change. This has served to foster an exclusivity that has pervaded the Commission, engaging only those whose expertise is sufficiently useful for the carbon city of Edinburgh.

On a personal level, our work on this project has hammered home how difficult it is to challenge and negotiate established, mainstream visions of the 'sustainable city'. This has been, at times, a very emotive and frustrating project which has tested our resilience and challenged our knowledge about climate governance in practice. However, despite the challenges described in this chapter, this hugely valuable professional and personal learning experience has highlighted the ways in which involvement in local climate action means wading into (and fully appreciating) the complex power relations involved with the governance of place-based futures. As the focus on place-based climate action intensifies, we hope that these reflections might provide some insightful lessons for others who find themselves in similar positions, at the confluence of urban climate governance in theory and in practice.

References

Angelo, H., & Wachsmuth, D. (2020). Why does everyone think cities can save the planet? *Urban Studies, 57*(11), 2201–2221.

Broto, V. (2019). Climate change politics and the urban contexts of messy governmentalities. *Territory, Politics, Governance*, 1–18.

Centre for Cities. (2019). *Cities Outlook 2019*. Available online: https://www.centreforcities.org/reader/cities-outlook-2019/a-decade-of-austerity/

De Volo, L. B., & Schatz, E. (2004). From the inside out: Ethnographic methods in political research. *Political Science and Politics, 37*(2), 267–271.

Evans, C., & Lambert, H. (2008). Implementing community interventions for HIV prevention: Insights from project ethnography. *Social Science & Medicine, 66*, 467–478.

Ford, M. (2019). *Nearly £1bn cut from Scottish authorities in past eight years.* Available online: https://www.localgov.co.uk/Nearly-1bn-cut-from-Scottish-authorities-in-past-eight-years/46830

Gramsci, A. (1971). *Selections from the prison notebooks.* International Publishers.

Hulme, M. (2019). Climate emergency politics is dangerous. *Issues in Science and Technology, 36*(1), 23–25.

Kenis, A., & Lievens, M. (2017). Imagining the carbon neutral city: The (post) politics of time and space. *Environment and Planning A: Economy and Space, 49*(8), 1762–1778.

Lewis, D., et al. (2003). Practice, Power and Meaning: Frameworks for studying organisational culture in multi-agency rural development projects. *Journal of International Development, 15,* 541–557.

Place-Based Climate Action Network. (2021). *About.* Available online: About PCAN | Place Based Climate Action Network (pcancities.org.uk).

Rice, J., Burke, B., & Heynen, N. (2015). Knowing climate change, embodying climate praxis: Experiential knowledge in Southern Appalachia. *Annals of the Association of American Geographers, 105*(2), 253–262. https://doi.org/10.1080/00045608.2014.985628

Stern, N. (2006). *Stern review on the economics of climate change (pre-publication edition). Executive Summary.* London: HM Treasury.

Swyngedouw, E. (2010). Apocalypse forever? Post-political populism and the spectre of climate change. *Theory, Culture & Society, 27*(2–3), 213–232.

Wolf, D. L. (2018). Situating feminist dilemmas in fieldwork. In *Feminist dilemmas in fieldwork* (pp. 1–55). Routledge.

Open Access This chapter is licensed under the terms of the Creative Commons Attribution 4.0 International License (http://creativecommons.org/licenses/by/4.0/), which permits use, sharing, adaptation, distribution and reproduction in any medium or format, as long as you give appropriate credit to the original author(s) and the source, provide a link to the Creative Commons license and indicate if changes were made.

The images or other third party material in this chapter are included in the chapter's Creative Commons license, unless indicated otherwise in a credit line to the material. If material is not included in the chapter's Creative Commons license and your intended use is not permitted by statutory regulation or exceeds the permitted use, you will need to obtain permission directly from the copyright holder.

CHAPTER 8

How Can 'Ordinary' Cities Become Climate Pioneers?

Wolfgang Haupt, Peter Eckersley, and Kristine Kern

Highlights We need to highlight the climate approaches of 'ordinary' cities, not just the high-profile leaders. 'Ordinary' cities can catch up with the leaders, even if they have only limited resources.

Keywords Climate adaptation · Climate mitigation · Climate governance · Pioneers · Local governance · 'Ordinary' cities

INTRODUCTION

Most academic studies into urban climate policies have focused on large forerunner cities, often highlighting how their ambitious and innovative approaches aim to deliver carbon neutrality by 2050 or earlier. Although such studies can tell a positive and inspiring story, these places often benefit from favourable conditions, such as higher levels of capacity or

W. Haupt (✉) · P. Eckersley · K. Kern
Leibniz-Institute for Research on Society and Space (IRS), Erkner, Germany
e-mail: wolfgang.haupt@leibniz-irs.de

© The Author(s) 2022
C. Howarth et al. (eds.), *Addressing the Climate Crisis*,
https://doi.org/10.1007/978-3-030-79739-3_8

community support for action. In addition, they only represent a small minority of the global population and an even smaller share of the world's cities. There are far more 'ordinary' cities—small and mid-sized municipalities, and places in the Global South as well as the Global North—than forerunners. To raise awareness of innovative practices that such places might wish to adopt, we need more research into how lower-profile cities are seeking to tackle climate change. This is because all local governments need to address climate change, and therefore, approaches have to be developed and shared that are applicable for a wide range of municipalities rather than just a handful of leaders.

Drawing on environmental and climate governance literature, this chapter explores and discusses the pathways such 'ordinary' cities might follow to become climate 'pioneers'. The chapter is sub-divided into four sections. Following this introduction, we introduce the idea of 'ordinary' cities and then explore how such cities might become climate pioneers, with specific reference to two cases from Germany. Finally, we summarise our findings and make some recommendations.

'Ordinary' Cities in a World of Global Cities

In the context of local climate action, we understand 'ordinary' cities as mostly mid-sized or smaller cities that are not high-profile progressive actors in climate governance. Wurzel et al. (2019) define climate *pioneers* as being less externally ambitious than climate *leaders*, and therefore, we might expect this term to be more applicable to ordinary cities than their global counterparts. Although lower-profile places may have pioneered innovative climate policies, they are generally not famous for having done so—perhaps because they have not developed particular city branding strategies or sought to position themselves as climate leaders (Gustavsson & Elander, 2012).

'Ordinary' cities can be best defined by identifying what they may lack: they benefit from neither a particular power of attraction, nor their

P. Eckersley
Nottingham Trent University, Nottingham, UK

K. Kern
Åbo Akademi University, Turku, Finland

extraordinary size or importance (Robinson 2020; Amin & Graham, 1997; Gerhard, 2017). Robinson (2002, p. 535) would refer to them as cities 'off the map'—at least in the eyes of most Western observers. She argues that scholars should seek to study wealthy and innovative cities alongside poorer cities in the Global South, to identify and exploit the opportunities to learn from a wide array of diverse urban contexts (Robinson, 2006). Given that the vast majority of cities have a much lower profile and are smaller in size than the handful of 'world cities' around the globe, we can see how the experiences of such 'ordinary' places are probably much more relevant for a wider range of urban areas. Therefore, if studies and practitioners focus predominantly on high-profile (often Anglophone) cities in the Global North, they are probably neglecting the innovations adopted elsewhere that may be much easier to apply in other contexts. As mentioned beforehand, all cities need to reduce greenhouse gas emissions and adapt to the impacts of climate change. We simply can no longer afford to ignore many of the (often very creative) approaches developed by 'ordinary' cities, regardless of their location.

From 'Ordinary' Cities to Climate Pioneers

According to Wurzel et al. (2019), potentially every city can become a pioneer; however, some cities are likely to find it easier than others. Indeed, previous research suggests that climate pioneers are typically characterised by high capacities for action (Haupt, 2020; Haupt et al., 2020; Homsy, 2018; Kern, 2019; Otto et al., 2021; Sharp et al., 2011) and a set of favourable socio-demographic, socio-economic and political conditions. These are: (i) a growing, young and above-average educated and skilled population (Bedsworth & Hanak, 2013; Kern, 2020; Zahran et al., 2008), (ii) favourable economic conditions such as high salaries (Bedsworth & Hanak, 2013; Kern, 2020; Zahran et al., 2008), (iii) support for climate action by city mayors (Bedsworth & Hanak, 2013; Haupt, 2020; Hoppe et al., 2016), (iv) political influence of green or alternative parties (Homsy, 2018; Mann et al., 2014), (v) a strong civil society (Homsy, 2018; Hoppe et al., 2016; Kern, 2019), particularly environmental groups (Sharp et al., 2011; Zahran et al., 2008), and (vi) a supportive local research environment (Eckersley, 2018; Kern, 2020).

Although 'ordinary' cities often lack many of these characteristics, the presence of powerful and committed actors within the municipality can help them to introduce pioneering initiatives (Pitt & Congreve, 2017). These individuals can be (a group of) policy entrepreneurs (Kingdon, 1984) such as specialised staff within the city government, or important key figures (Gailing & Ibert, 2016) such as city mayors. Climate action does not necessarily need to be initiated by the mayor or a leading politician, but local policy entrepreneurs who wish to introduce ambitious policies will need their political support (Young, 2010). To promote policy innovations, policy entrepreneurs also need to dedicate a significant amount of their own resources (e.g. time, capabilities and reputation) and identify the 'right moment' (a 'policy window') to place their topic on the agenda (Mintrom, 2019).

CHANCES AND CHALLENGES FOR 'ORDINARY' CITIES'

This section explores how two 'ordinary' cities in Germany: Göttingen (120.000 inhabitants) and Remscheid (110.000 inhabitants) became climate pioneers. Our discussion of Göttingen draws heavily on Fenton and Paschek's (2018) study, which was based on five expert interviews and an analysis of key strategic documents. We undertook 11 fieldwork interviews in Remscheid ourselves, and also conducted extensive analysis of relevant documentation (see Haupt & Kern, 2020 for a more detailed examination of this case).

Neither city has the characteristics that we would normally associate with climate leaders. Two extensive studies that investigated a broad set of socio-economic, socio-demographic and socio-cultural indicators in all German cities and counties (*Landkreise*) found that Remscheid ranks 360th out of 401 in terms of overall quality of life[1] and 323rd in prospected future opportunities,[2] whereas Göttingen was placed 158th in both studies. Göttingen does exhibit some features of a typical pioneer: it is a university city with a young and well-educated population. Nevertheless, an examination of the city's climate action revealed a high degree of capacity deficits and a severe lack of municipal resources (Fenton & Paschek, 2018). For its part, the shrinking industrial city of Remscheid,

---

[1] *ZDF-Die große Deutschland-Studie 2019:* the study analysing the overall quality of life (https://deutschland-studie.zdf.de/district/09162/default).

[2] *Prognos Zukunftsatlas 2019:* the study analysing prospected future opportunities *Prognos Zukunftsatlas* (https://www.handelsblatt.com/politik/deutschland/zukunftsatlas-2019/).

with its very high municipal debt and the resulting capacity constraints, is a textbook example of a likely laggard (Haupt & Kern, 2020). Despite these disadvantages, both places took action earlier than most other German cities of comparable size and completed a wide array of climate activities (see Otto et al., 2021). Table 8.1 summarises several key milestones Göttingen's and Remscheid's climate action activities.

To cope with their rather unfavourable environments for pioneering climate policies, both cities needed to look for alternative creative approaches. Most importantly, Göttingen and Remscheid collaborated with local key actors (such as universities) to bid successfully for third-party funding and participate in temporary projects (Fenton & Paschek, 2018; Haupt & Kern, 2020). This has given Remscheid access to crucial knowledge for developing its climate policy (e.g. in the creation of local climate analysis maps and maps for simulating flow pathways and depressions in the case of heavy rainfall). Indeed, all of the city's climate-related strategies (mitigation, adaptation, mobility) were developed as

Table 8.1 Milestones of local climate action (external funding bodies in brackets where applicable)

Göttingen	*Remscheid*
• 1990: energy strategy	• 1995: entry into the Climate Alliance
• 1991: entry into the Climate Alliance	• 1998: ratification of the Alborg Charta, Local Agenda 21 resolution
• 1997: Local Agenda 21 resolution	• 1999: climate mitigation strategy (*federal state funding*)
• 2010: climate mitigation strategy (*national funding*)	• 2003, 2007, and 2018: European Energy Award certification (*federal state funding*)
• 2014: climate strategy aiming at climate neutrality by 2050 (*national funding*)	• Climate manager appointed in the department of building management (*national funding*)
• 2017: climate manager appointed (*national funding*)	• 2013: climate adaptation strategy (*national funding*)
• 2017: mobility strategy (*national funding)*	• 2014: integrated climate mitigation strategy (*national funding*)
• 2018: bicycle traffic development strategy	• 2017: climate manager appointed (*national funding*)
	• 2018: mobility strategy (*national funding*)

Source Own table

part of funding programmes (Haupt & Kern, 2020). Similarly, Göttingen's mitigation strategies and climate traffic plan would not have come about without external funding (Fenton & Paschek, 2018). Moreover, the city succeeded in a competitive tender process to participate in the 'Master plan 100% Climate Protection' funding programme that financially supports the development of mitigation plans aiming at climate neutrality by the year 2050.

Nevertheless, dependency on third-party funds can lead to uncertainties in mid-term and long-term planning, because externally-funded initiatives are often difficult to sustain after projects are completed. Conscious of this risk, Göttingen has sought to increase public participation in decision-making and policy formulation, in the hope that this will create capacity within the city to maintain momentum (Fenton & Paschek, 2018). In Remscheid, the reliance on external funding had a detrimental impact on the setting of ambitious long-term climate goals and developing holistic visions for the future (Haupt & Kern, 2020). Although Göttingen had already managed to set very ambitious targets, it still faced the challenge of implementing its strategy and requires ongoing resources to achieve its climate objectives (Fenton & Paschek, 2018). Nevertheless, external funding can also lead to the implementation of concrete measures. As an example, both cities received national grants to fund the (temporary) employment of a climate manager. Further examples from Remscheid include various energy-saving projects in public schools (funded by the German Federal Ministry for the Environment) and building up a green facade at a school building (funded by the Federal Ministry of Education and Research) (Haupt & Kern, 2020).

Outlook and Recommendations

The previous section highlighted how third-party funding for climate-related projects can help 'ordinary' cities to become pioneers. The German mid-sized cities of Göttingen and Remscheid show that strong key actors that manage to attract external funding from a variety of sources can—to a certain extent—compensate for a lack of capacities and resources. External funding can enable the development of strategies, engagement of temporary staff and also—to a lesser degree—the implementation of mitigation or adaptation measures. However, it is as yet unclear as to whether such an approach provides the only feasible opportunity for 'ordinary' or even 'disadvantaged' cities to become climate pioneers—and questions remain as to whether reliance on third-party funding is an effective strategy over the longer term, given that projects often cease when the money runs out. It would hardly be desirable if bidding for such funding represents the only creative approach that 'ordinary' cities can pursue to advance local climate policymaking, but the lack of studies into this area means that we do not know enough about the other strategies that they may have adopted to tackle climate change.

As we have discussed, focusing on large leading forerunner cities is problematic: first, because there is only a limited number of such cities, and second because their models and solutions are barely replicable for the vast majority of 'ordinary' cities that operate within very different contexts. Indeed, cities that have pursued approaches that are more likely to be transferable between 'ordinary' cities should receive more attention. Shining a spotlight on their pioneering activities might not make these places *extra*ordinary, but it can help to raise awareness of the types of climate initiatives or place-based approaches that other 'ordinary' cities might consider adopting. More studies into these urban areas should help to highlight and spread the word about innovative practices in under-researched places. In addition, municipal practitioners could involve themselves in city networks and exchanges to learn more about how the pioneering approaches of other places might be applicable to their own contexts.

There is growing evidence that there are numerous undetected and unrecognised 'ordinary' cities out there that, despite a lack of attention, have the potential to develop creative and pioneering approaches, successfully tackle climate change issues within city borders and catch up to the leaders. Many of them have probably developed diverse creative approaches already, but have not received due credit for their efforts and remain largely unacknowledged. Indeed, it is most likely that Göttingen and Remscheid are not the only German examples of such cities. Ideally, their creative approaches could serve as inspiration or even models for the numerous cities that are in a similar situation and thereby help many other 'ordinary' or even 'disadvantaged' cities to develop successful approaches themselves.

References

Amin, A., & Graham, S. (1997). The ordinary city. *Transactions of the Institute of British Geographers, 22*(4), 411–429.

Bedsworth, L. W., & Hanak, E. (2013). Climate policy at the local level: Insights from California. *Global Environmental Change, 23*(3), 664–677.

Biesbroek, G. R., et al. (2013). On the nature of barriers to climate change adaptation. *Regional Environmental Change, 13*(5), 1119–1129.

Castán Broto, V. (2020). Climate change politics and the urban contexts of messy governmentalities. *Territory, Politics, Governance, 9*(2), 241–258.

Eckersley, P. (2018). *Power and capacity in urban climate governance: Germany and England compared.* Peter Lang.

Fenton, P., & Paschek, F. (2018). Projects, participation and planning across boundaries in Göttingen. *Regional Studies, Regional Science, 5*(1), 81–89.

Gailing, L., & Ibert, O. (2016). Schlüsselfiguren: Raum als Gegenstand und Ressource des Wandels. *Raumforschung Und Raumordnung, 74*(5), 391–403.

Gerhard, U. (2017). *Mega city, global city, ordinary city - Zeitgemäße Begriffe einer kosmopolitanen, interdisziplinären Stadtforschung?* Marsilius-Kolleg.

Gustavsson, E., & Elander, I. (2012). Cocky and climate smart? Climate change mitigation and place-branding in three Swedish towns. *Local Environment, 17*(8), 769–782.

Haupt, W. (2020). How do local policy makers learn about climate change adaptation policies? Examining study visits as an instrument of policy learning in the European Union. *Urban Affairs Review.*

Haupt, W., et al. (2020). City-to-city learning within climate city networks: Definition, significance, and challenges from a global perspective. *International Journal of Urban Sustainable Development, 12*(2), 143–159.

Haupt, W., & Kern, K. (2020). *Entwicklungspfade von Klimaschutz und Klimaanpassung in Remscheid.* Erkner.

Homsy, G. C. (2018). Unlikely pioneers: Creative climate change policymaking in smaller U.S. cities. *Journal of Environmental Studies and Sciences, 8*(2), 121–131.

Hoppe, T., van der Vegt, A., & Stegmaier, P. (2016). Presenting a framework to analyze local climate policy and action in small and medium-sized cities. *Sustainability, 8*(9), 847.

Kern, K. (2019). Cities as leaders in EU multilevel climate governance: Embedded upscaling of local experiments in Europe. *Environmental Politics, 28*(1), 125–145.

Kern, K. (2020). Von Vorreitern und Nachzüglern: Die Rolle von Städten und Gemeinden in der Klimapolitik. In T. Hickmann & M. Lederer (Eds.), *Leidenschaft und Augenmaß: Sozialwissenschaftliche Perspektiven auf Entwicklung, Verwaltung, Umwelt und Klima* (pp. 195–206). Nomos.

Kingdon, J. W. (1984). *Agendas, alternatives and public policies.* HarperCollins.

Mann, S., Briant, R. M., & Gibin, M. (2014). Spatial determinants of local government action on climate change: An analysis of local authorities in England. *Local Environment, 19*(8), 837–867.

Mintrom, M. (2019). *Policy entrepreneurs and dynamic change.* Cambridge University Press.

Otto, et al. (2021). https://doi.org/10.1007/s10584-021-03142-9

Pitt, D., & Congreve, A. (2017). Collaborative approaches to local climate change and clean energy initiatives in the USA and England. *Local Environment, 22*(9), 1124–1141.

Robinson, J. (2002). Global and world cities: A view from off the map. *International Journal of Urban and Regional Research, 26*(3), 531–554.

Robinson, J. (2006). *Ordinary cities: Between modernity and development.* Questioning Cities Series. Routledge Taylor & Francis Group.

Robinson, J. (2020). World cities, or a world of ordinary cities? In R. T. LeGates & F. Stout (Eds.), *The city reader* (7th ed., pp. 678–688). Routledge.

Sharp, E. B., Daley, D. M., & Lynch, M. S. (2011). Understanding local adoption and implementation of climate change mitigation policy. *Urban Affairs Review, 47*(3), 433–457.

Wurzel, R. K. W., Liefferink, D., & Torney, D. (2019). Pioneers, leaders and followers in multilevel and polycentric climate governance. *Environmental Politics, 28*(1), 1–21.

Young, R. F. (2010). The greening of Chicago: Environmental leaders and organisational learning in the transition toward a sustainable metropolitan region. *Journal of Environmental Planning and Management, 53*(8), 1051–1068.

Zahran, S., et al. (2008). Vulnerability and capacity: Explaining local commitment to climate-change policy. *Environment and Planning C: Government and Policy, 26*(3), 544–562.

Open Access This chapter is licensed under the terms of the Creative Commons Attribution 4.0 International License (http://creativecommons.org/licenses/by/4.0/), which permits use, sharing, adaptation, distribution and reproduction in any medium or format, as long as you give appropriate credit to the original author(s) and the source, provide a link to the Creative Commons license and indicate if changes were made.

The images or other third party material in this chapter are included in the chapter's Creative Commons license, unless indicated otherwise in a credit line to the material. If material is not included in the chapter's Creative Commons license and your intended use is not permitted by statutory regulation or exceeds the permitted use, you will need to obtain permission directly from the copyright holder.

PART III

The Agents of Local Climate Action

Effective Communication on Local Adaptation: Considerations for Providers of Climate Change Advice and Support

Kristen Guida and Candice Howarth

Highlights There is a reduction in adaptation action despite science continuing to produce actionable material. Climate change adaptation advice and support must be salient, credible and legitimate among other things.

Keywords Climate adaptation · Science-practice interface · Climate knowledge · Local action · Stakeholder engagement

K. Guida
London Climate Change Partnership, London, UK
e-mail: kristen.guida@london.gov.uk

C. Howarth (✉)
Grantham Research Institute on Climate Change and the Environment,
London School of Economics and Political Science, London, UK
e-mail: c.howarth@lse.ac.uk

© The Author(s) 2022
C. Howarth et al. (eds.), *Addressing the Climate Crisis*,
https://doi.org/10.1007/978-3-030-79739-3_9

Introduction: Ensuring User Uptake of Climate Science

Climate change is a global issue with impacts felt at the local level, where many solutions to climate risks are implemented (Howarth et al., 2021). Working at the climate adaptation, science-practice interface requires interaction and collaboration across scales, sectors and stakeholders to identify climate risks, potential solutions and decision points that provide opportunities to enhance adaptation. However, with scarce resources, skills and capacity, resilience and adaptation are seldom embedded in local climate action plans. In part, overcoming barriers that occur at the science-practice interface requires careful consideration of the credibility, salience and legitimacy of scientific knowledge.

This chapter explores the challenges that emerge in the science-practice interface and the extent to which the translation of climate science into action is enhanced or inhibited when it comes to adaptation action and building resilience. Through a case study exploring efforts to bridge the gap between national climate risk assessments and adaptation planning, we analyse how the balance of salience, credibility and legitimacy of knowledge is considered at the science-practice interface and discuss implications of this for local climate adaptation. There are a number of ways to examine the science-practice gap and to consider the barriers to uptake and implementation of robust adaptation processes. With calls for clearer and more useful scientific information (McNie, 2007), this gap is often presented as a communications problem, with scientists urged to communicate more effectively, simplify findings, engage proactively with users and find innovative routes to specific audiences (Bidwell et al., 2013; Hine et al., 2014; Lemos et al., 2012). However, scientists are often not trained in the art of science communication and meaningful engagement that enables mutual understanding of 'real-world' needs and the ability of science to meet them. In addition, potential users of scientific information often lack sufficient understanding about their adaptation evidence needs to communicate these effectively to scientists.

We need to understand the barriers and issues for decision-makers whose evidence requirements may vary widely, and there is a growing need to understand how to translate science more effectively into practice to inform decision-making at different scales. In this chapter, we explore how this translation can happen effectively, and in particular, we explore whether our established way of thinking about the science-practice gap (placing the burden of communication on scientists) is helpful. Or alternatively, whether we need to rethink the emphasis and focus to enable a productive two-way conversation between science and practice.

THE EVOLVING LANDSCAPE OF LOCAL CLIMATE ADAPTATION IN THE UK

The recognition of the importance of climate change adaptation in the UK gathered momentum around the turn of the twenty-first century. From the outset, the need for science-to-practice engagement and communications to support local adaptation was recognised, and regional climate change partnerships (CCPs) were established in England to gather evidence and support risk assessments and adaptation by local authorities, businesses, communities and other actors. In 2008, the UK Climate Change Act enshrined adaptation in law, and from 2008 to 2010, local authorities were required to report against a national indicator (NI 188) on their preparations for a changing climate. To support this effort, the CCPs were given resource by government to boost coordination of knowledge sharing and collaboration. They worked closely with the UK Climate Impacts Programme (UKCIP), and later with the Environment Agency's Climate Ready support service, to turn the science into decision-support tools and resources to tailor advice and support according to specific audiences, ensuring research met real-world needs rather than just filling academic knowledge gaps. Examples of this activity include the dissemination and development of tools like UKCIP's Business Areas Climate Assessment Tool (BACLIAT) and Local Climate Impacts Profile (LCLIP); a Business Resilience Health Check; regional-level sector impacts studies and sector-specific adaptation guides; and national programmes of training and engagement for everyone from health sector practitioners to highways officers and planners—including a nationally-accredited qualification on business resilience. CCPs also worked with local and regional stakeholders to produce regional 'translations' of the first UK Climate Change Risk Assessment in 2012 and supported the production of local risk assessments informed by both of the national assessments.

The guiding principle behind all work on local climate adaptation was that it had to reflect the needs and priorities of the decision-makers and practitioners who would use it. A tool or resource might be generic, but it could be presented in a way that would be meaningful for the audience and applicable to the practitioner's role or task. This meant a fair amount of stakeholder engagement—including relationship building, translation and understanding of context and priorities—that researchers do not often have the time or skill to achieve. However, since the withdrawal of NI 188 and the disbanding of most CCPs a few years later, there is a gap in the supply of locally-relevant evidence and the coordination of stakeholder engagement for awareness and capacity building. In recent years, the support for engagement and science-to-practice coordination has disappeared, but the need for it remains as urgent as ever. The UK's Climate Change Committee (CCC) has recognised this in their reports on the UK's progress on adaptation (CCC, 2021, 2019a). In preparation for the third UK Climate Change Risk Assessment (CCRA), the CCC commissioned a team led by current and former CCP coordinators to conduct a study of how to improve the CCRA's accessibility—and thus ensure that the National Adaptation Programme effectively addressed the country's main climate risks.

Case Study: Assessing Accessibility of the UK's Climate Change Risk Assessment

The Climate Change Committee (CCC) is an independent statutory body established under the 2008 Climate Change Act to advise the UK and devolved governments on emissions targets and to report to Parliament on progress towards mitigating and adapting to climate change. In 2019, the CCC's Adaptation Committee commissioned a consortium led by Sustainability West Midlands (SWM) to lead a project to improve the accessibility of the UK third Climate Change Risk Assessment Evidence Report, due to be published in 2021 (CCC, 2019b). The aim of the project was also to 'provide the CCC with advice and products to improve the impact of the CCRA by enhancing its accessibility to its primary customers, which are UK Government departments, the devolved administrations and government-funded arm's length bodies' (p. 3). Whilst local government is not considered a primary customer, the CCC's efforts to improve the take-up of scientific evidence provide useful learning for efforts to promote adaptation at the local level. Evidence has shown that the first two CCRAs, published in 2012 and 2017, whilst demonstrating improvements, were difficult for government officials and stakeholders to access and use effectively (Howarth & Painter, 2016; Howarth et al., 2018). The third CCRA, as with the previous two, will inform the subsequent National Adaptation Programme (NAP) produced by the Department for Food, Environment and Rural Affairs in 2023. Accessibility of the CCRA's products, an understanding of the context in which these will be received and used, is fundamental to an effective NAP and to building the UK's resilience to climate change impacts.

The project delivering this work is ongoing, and hence, we have focused our (thematic) analysis on the tender document used to commission it (see CCC, 2019b). We acknowledge therefore that the delivery of this work may differ from what was set out in the tender; nevertheless, we see this as a useful way to explore how the accessibility strategy of the CCRA3 Evidence Report was informed by the guidance provided for the commissioning of this research and whether this considered aspects of salience, credibility and legitimacy in the project aims.

We conducted this analysis using Cash et al. (2002)'s framework on credibility, salience and legitimacy. It suggests that the boundary between science and policy or science and practice is one of the barriers to effective uptake of scientific knowledge to inform decision-making. In order to overcome this, Cash et al. argue that evidence used to inform decision-making must be *credible* (e.g. authoritative, believable, trusted), *salient* (information relevant to decision-makers' decisions) and *legitimate* (information produced is unbiased and fair and considers the values and needs of different actors). The challenge with this, however, is that actors on either side of the science-practice boundary see and value credibility, salience and legitimacy differently. Consequently, in order for scientific knowledge to be taken up and connected to action, efforts to facilitate this must be simultaneously credible, salient and legitimate to all stakeholders involved (Kunseler et al., 2015). However, often credibility is the predominant focus and salience and legitimacy are given different, lesser weights and efforts to address one can enhance or dampen the efficacy and focus of the other attributes. For example, efforts to give more prominence to legitimacy of information can affect (negatively or positively) the extent to which it is salient to the audience in question. This framework has been used to explore similar processes assessing the usability and accessibility of climate change evidence (Howarth & Painter, 2016) and provides a useful and user-friendly way of analysing the CCC's tender document for its CCRA3 accessibility project.

Results of our analysis, using Cash et al.'s (2002) framework on salience, credibility and legitimacy criteria, are presented in Table 9.1. *Salience* of CCRA3 is addressed throughout the tender document with a

Table 9.1 Considerations for Cash et al.'s (2002) salience, credibility and legitimacy criteria in the CCC's tender documentation (CCC, 2019b)

	Example reference in tender document
Salience For example, relevant to decision-makers' decisions	'Both Government and the CCC are therefore interested in understanding how CCRA2 was used by its primary customer group (UK Government departments, devolved administrations and arm's length bodies), what the barriers were to obtaining the information required for different stakeholders to develop their plans, and how the summary materials should therefore be produced for CCRA3 to assist in improving its usability and impact as a resource for Government' (Background section, p. 4) 'The aim of this project is to help the CCC to present the CCRA3 Evidence Report in such a way as to best enable the Government, devolved administrations and arm's length bodies to use the outputs effectively in their resulting national adaptation programmes' (Aims and Objectives, p. 4)
Credibility For example, authoritative, believable, trusted	'It is important for the CCRA to be carried out in a robust, independent and transparent way, and for all supporting evidence, assumptions and rationale to be provided' (Background section, p. 3) 'The CCC has limited budget/staff to undertake a major communications campaign in-house, and contractors will need to be mindful of this when formulating their recommendations' (Challenges, p. 10) 'All applicants will need to identify and propose arrangements for initial scrutiny and ongoing monitoring of ethical issues. The appropriate handling of ethical issues is part of the tender assessment exercise and proposals will be evaluated on this as part of the 'addressing challenges and risks' criterion' (Ethics, p. 11)

(continued)

Table 9.1 (continued)

	Example reference in tender document
Legitimacy For example, unbiased, fair and considers values and needs of different actors	'Getting to the heart of the accessibility issue for CCRAs; this will include taking into account differing perspectives from different stakeholders and judging what outcomes would work best to improve the impact of CCRA3 for its main customer. Considerable knowledge and experience of communications best practice will be required' (Challenges, p. 10) 'Understanding the perspectives of stakeholders sufficiently to work out what they need. This may be very simple materials rather than more innovative forms of communication. **User needs must come before the desire for innovation or being 'on trend' in this project**' (Challenges, p. 10, text in bold emphasised in document)

focus on the use of the report by its primary customer group (i.e. government) and the extent to which lessons on how previous CCRAs were used to inform decision-making could help improve usability and impact of CCRA3. This is specifically addressed by capturing views of government officials and other stakeholders (business is given as an example) of barriers to uptake and use of CCRAs 1 and 2 Evidence Reports and supporting materials such as charts, diagrams and descriptions of the risks and opportunities. There is, however, no explicit mention of consulting stakeholders (government and others) as to what specifically they require to enable the effective use of these materials and any institutional, capacity, resource or knowledge barriers that will affect CCRA3 uptake. Nevertheless, stakeholders were consulted throughout the project via surveys, interviews and workshop sessions about what barriers and enablers they saw to the effective use of the CCRA materials. In addition, the selection of project outputs analyses (e.g. sector fact sheets, country summaries, new CCRA website) also supports a strong emphasis on salience.

The *credibility* of the work needed is emphasised strongly from the outset with a focus on robustness, independence and transparency. A strong emphasis on learning from wider contexts emerges as does quality assurance, with open and regular communication between the awarded project consortium and the CCC. The robustness of the project is further enhanced by engagement between the consortium and other CCC-funded CCRA3-relevant activities to ensure consistency, alignment and ultimately trust in the outcomes and outputs. With a major output of the work being a communication strategy for CCRA3, the tender highlights the limitations of the CCC in-house team in terms of capacity, resource and skills, to deliver such strategy. Whilst this may be seen as calling into question the ability to deliver the recommendations set out by the project, we consider this to be a positive acknowledgement of existing gaps and a constructive way to ensure the delivery of the work takes this into consideration.

The *legitimacy* of the work is prominently reflected in the project. Acknowledging limitations of previous CCRAs, the awarded consortium is expected to take seriously previous concerns as well as the needs and views of end-users to improve take-up of the findings of CCRA3. This speaks strongly to Cash et al.'s legitimacy criteria and prioritises this over a 'desire for innovation or being 'on trend' in this project' (p. 10). This is a requirement both in the outputs produced and in the method of working with government officials.

Credibility, salience and legitimacy are fairly well considered in the tender document and help shape and define what is expected. However, there is little to indicate how the project views the end result of the effort; that is, what success looks like and how this will address issues discussed above in regard to capacity of end-users to deliver on the recommendations made. In particular, a clear formulation of what the issue is, beyond simply improving communication and instead outlining clearly what issues exist for different decision-makers across different policy areas, levels and sectors, requires a sustained monitoring effort.

CONCLUDING REMARKS

By exploring the CCC's efforts to improve accessibility and uptake of the CCRA3 in the light of the salience, legitimacy and credibility framework, we seek to encourage more critical consideration and analysis of the way in which climate change science is taken up in practice to inform adaptation measures and resilience on the ground, including at the local level. We note, however, that communication alone is not the sole issue to consider, and that the results of this accessibility project will not be known for a while. We also note that the main audience identified by the CCC for both this project and the CCRA3 is central government, despite an awareness that CCRA3 outputs will also be used to inform local adaptation and resilience.

The barriers to uptake considered as part of the CCRA3 accessibility project referred mainly to ongoing developments in the production of the CCRA3 visual and written outputs and government timelines, with little mention of the pressures and processes that may affect use, usability and accessibility of the CCRA3 products by end-users (and end-users beyond the primary customer group, particularly at the local level). Even where the best science and communication are in place, other barriers may exist. Organisations often lack critical capacity in terms of personnel within organisations to use the science (e.g. often there is nobody with climate change adaptation in their job description), and they may lack knowledge and skills to use the science (i.e. adaptation requires a different set of skills from many other environment-/climate-related roles like carbon mitigation or waste/circular economy) and of how climate change relates to a particular organisation's objectives and priorities; there is the perception of adaptation as an environmental issue with lesser importance; a lack of support from leadership can manifest in a lack of urgency and resources assigned to adaptation, and regulatory frameworks can also inhibit consideration of climate change or longer-term risks. Similarly, other policies and priorities often compete or conflict with climate change adaptation. Finally, the emphasis on a 'science first' approach to adaptation rather than a 'context-based' approach can lead to unnecessary and distracting confusion over the uncertainties of climate science.

The accessibility project meets the credibility, legitimacy and salience tests as set out by Cash et al., and as such could stand as a good example of how to consider or commission communication in the science-practice space. However, the results of this work remain to be seen and evaluated. The authors suggest that any future evaluation considers the role of other, non-communication-related barriers to uptake, namely capacity and the lack of an enabling environment.

Often the provision of science and evidence to inform practice is explored as are the gaps, challenges and barriers to 'using' scientific evidence. What is needed is a better and deeper understanding of where the relationship between science and practice breaks down and how this manifests at a local level where many climate adaptation solutions are implemented. How science is communicated is only one of many challenges; there is also a need to recognise the range of other factors that act as barriers to take up.

References

Bidwell, D., Dietz, T., & Scavia, D. (2013). Fostering knowledge networks for climate adaptation. *Nature Climate Change, 3*, 610–611.

Cash, D. W., Clark, W., Alcock, F., Dickson, N. M., Eckley, N., & Jaeger, J. (2002). *Salience, credibility, legitimacy and boundaries: Linking research, assessment and decision-making* (KSG Working Paper Series RWP02-046). Harvard University, Cambridge, MA.

Committee on Climate Change. (2019a). *Progress in preparing for climate change—2019 Progress Report to Parliament.* CCC.

Committee on Climate Change. (2019b). *Invitation to tender for research: Improving accessibility of the UK Climate Change Risk Assessment Evidence Report* (Tender Ref. KB-0819).

CCC. (2021). *Progress in adapting to climate change—2021 Report to Parliament.* https://www.theccc.org.uk/publication/2021-progress-report-to-parliament/.

Hine, D. W., Reser, J. P., Morrison, M., Phillips, W. J., Nunn, P., & Cooksey, R. (2014). *Audience segmentation and climate change communication: Conceptual and methodological consideration* (WIRES Climate Change).

Howarth, C., Barry, J., Fankhauser, S., Gouldson, A., Lock, K., Owen, A., & Robins, N. (2021). *Trends in local climate action in the UK.* A report by the Place-Based Climate Action Network (PCAN), UK.

Howarth, C., Morse-Jones, S., Brooks, K., & Kythreotis, A. (2018). Co-producing UK climate change adaptation policy: An analysis of the 2012 and 2017 UK Climate Change Risk Assessments. *Environmental Science and Policy, 89*, 412–420.

Howarth, C., & Painter, J. (2016). The IPCC and local decision-making on climate change: A robust science-policy interface? *Palgrave Communications, 2*, 16058.

Kunseler, E.-M., Tuinstra, W., Vasileiadou, E., & Petersen, A. C. (2015). The reflective futures practitioner: Balancing salience, credibility and legitimacy in generating foresight knowledge with stakeholders. *Futures, 66*, 1–12. https://doi.org/10.1016/j.futures.2014.10.006

Lemos, M. C., Kirchhoff, C. J., & Ramprasad, V. (2012). Narrowing the climate information usability gap. *Nature Climate Change, 2*, 789–794.

McNie, E. C. (2007). Reconciling the supply of scientific information with user demands: An analysis of the problem and review of the literature. *Environmental Science & Policy, 10*(1), 17–38. https://doi.org/10.1016/j.envsci.2006.10.004

Open Access This chapter is licensed under the terms of the Creative Commons Attribution 4.0 International License (http://creativecommons.org/licenses/by/4.0/), which permits use, sharing, adaptation, distribution and reproduction in any medium or format, as long as you give appropriate credit to the original author(s) and the source, provide a link to the Creative Commons license and indicate if changes were made.

The images or other third party material in this chapter are included in the chapter's Creative Commons license, unless indicated otherwise in a credit line to the material. If material is not included in the chapter's Creative Commons license and your intended use is not permitted by statutory regulation or exceeds the permitted use, you will need to obtain permission directly from the copyright holder.

CHAPTER 10

Diversifying the Private Sector in Local Climate Commissions

Robert Connell and Matthew Lane

Highlights We ask who exactly it is that represents the 'place'-based interests of the private sector? Nuanced understanding of private sector required; beyond simply the biggest emitters and richest organisations.

Keywords Climate commissions · Place-based · Private sector · Climate action

INTRODUCTION

In this chapter, we reflect on desk-based research carried out to support the Place-based Climate Action Network's (PCAN) project's ambitions to establish impactful climate commissions in cities across the UK (PCAN,

R. Connell (✉)
University of Edinburgh, Edinburgh, Scotland, UK

M. Lane
School of Geosciences, University of Edinburgh, Edinburgh, Scotland, UK
e-mail: matthew.lane@ed.ac.uk

© The Author(s) 2022
C. Howarth et al. (eds.), *Addressing the Climate Crisis*,
https://doi.org/10.1007/978-3-030-79739-3_10

2020). These aim to act as an independent body of place-based stakeholders (situated actors invested in the future of a location) working to accelerate action on climate change in their 'place'. Using the current compositions of the three PCAN Climate Commissions (Belfast, Edinburgh and Leeds) as our starting point, in this chapter, we engage with the question of how existing understandings of the private sector might be broadened to more holistically capture the diverse perspectives and opportunities that the sector might offer to the work of city climate commissions.

Researchers on the subject of climate change and business often tout the private sector as the vehicle through which to realise ambitious climate targets via economic incentives and regulation (Averchenkova et al., 2020; Stern, 2007). It is a robust argument to state that the private sector is a powerhouse of economic, logistical and innovative momentum which (in theory) has the capacity to act on climate change, if appropriately harnessed (Hawken, 2017). As such, and acknowledging the private sector's omnipresence in industrialised society, city-based climate commissions have identified the private sector as a potential source of influential individuals and organisations (PCAN, 2020). The question of who gets a seat at the table, however, will be place-specific and therefore arguably in need of a more nuanced approach than simply championing the involvement of the 'private sector'. Indeed, we ask here, who exactly is it that represents the 'place'-based interests of the private sector?

Reflecting on our critical and conceptual engagement with this question, we suggest here that climate commissions approach the private sector with a degree of granularity, rather than perceiving it as a collection of independent businesses clustered into one spatial monolith. To encourage more effective fostering of local climate action on the part of the business environment, we hope to describe some of the benefits that climate commissions could gain by understanding and integrating a multiplicity of a city's stakeholders from the wider private sector 'ecosystem'. Including a more eclectic range of local actors in place-based climate commissions would offer new perspectives, knowledge and influence to draw from. In this chapter, we identify and discuss some specific examples of place-based stakeholders from the private sector, that have thus far been overlooked by PCAN commissioner recruitment and reflect on the insight they might provide in enhancing local climate action.

In the following section, we present the output of a research project which sought to distil, from existing academic literature, a comprehensive framework for capturing the diverse and varying set of competitive and economic benefits, or motivators, for private sector organisations to take action against climate change. Whilst this framework yields numerous theoretical insights and further avenues of research, its use in this chapter is practical—as a device to identify new actors with something to offer to the ambitions of climate commissions. In the subsequent sections of the chapter, we connect the identified private sector motivators for climate action from the framework to the sorts of stakeholders of a locality that are able to speak to these motivations at the place-based scale. In doing so, the framework enables the identification of (and offers recommendations on) actors who might be considered when deciding on the composition of place-based climate commissions. This is based on their capacity to mobilise different agencies with local private sector ecosystems. In conclusion, we offer some further (more speculative) thoughts on the relationship between climate action, place and the private sector.

Mapping Business Motivations for Climate Action

The research project on which this chapter is based took the form of a comprehensive desk-based, 'Integrated literature review' (Torraco, 2005) carried out by the chapter's lead author between April and August 2020.[1] The motivation to create a synthesised conceptual framework emerged out of a necessity to tie together a largely disparate body of private sector and climate action literature in order to move debates beyond the 'usual suspects' of the field (Evans, et al., 2017; Hoffman, 2016; Schaltegger, et al., 2012). Figure 10.1 captures the output of this process as a new framework for approaching the topic of private sector motivations for taking action against climate change.

Each individual motivator presented in Fig. 10.1 thematically captures a cluster of economic and strategic benefits for private sector action on climate change from within the academic literature. Having grouped these into 5 conceptual clusters (or 'lenses'), and beginning with explicitly financial motivations, in what follows we connect the identified clusters to potential actors who (a) have the potential to offer much needed insight

---

[1] This study formed part of dissertation research on the MSc Carbon Management at the University of Edinburgh.

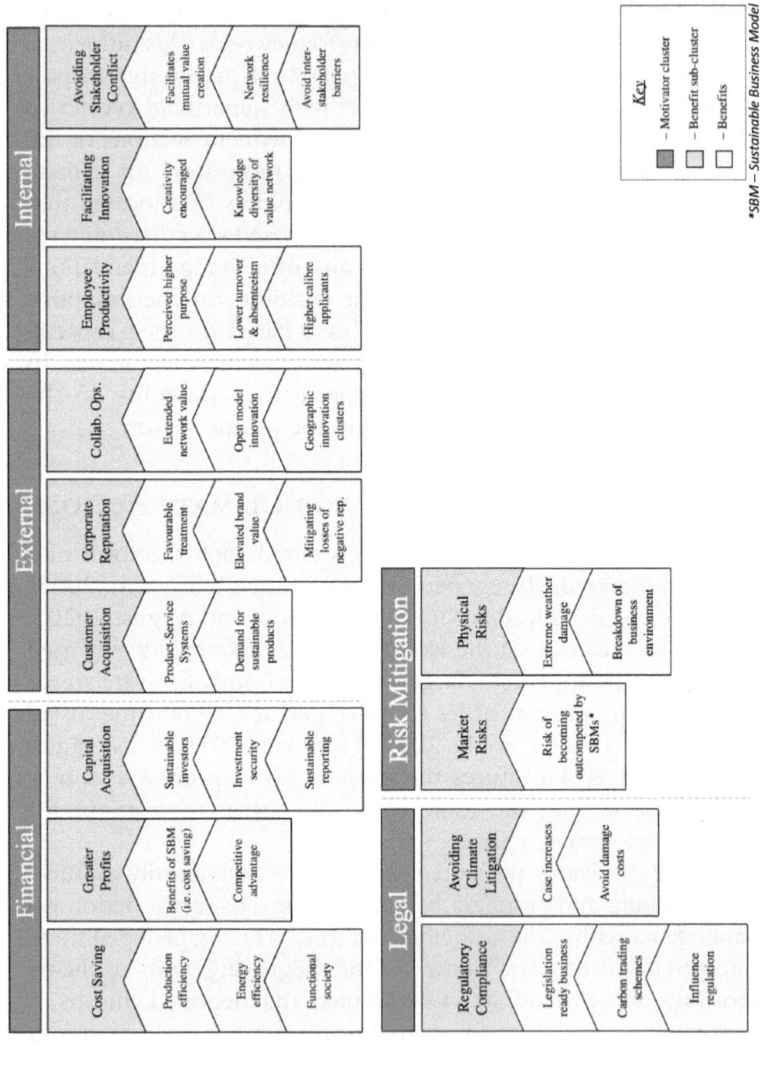

Fig. 10.1 A framework of motivators for private sector action on climate change (*SBM—Sustainable Business Model)

into the nature of this motivator in a particular city or place, and (b) were overlooked during initial recruitment to, or have not yet been adopted throughout, PCAN Climate Commissions.

FINANCIAL LENS: CHAMBER OF COMMERCE

Despite a noteworthy presence of numerous individual business memberships in PCAN Climate Commissions, there is yet to be a representative whose remit encompasses the totality of the local business ecosystem. Whilst such a perspective might not exist with regard to the question of climate change, it is possible to identify organisations and groups who *do* have an explicit relationship with the 'place-based' dimension of business activity. Chambers of Commerce, for example, offer such a perspective, acting as an advocate for the interests of large quantities of businesses in very particular places (Chamber of Commerce, 2020). Rather than merely being the local manifestation of otherwise national or even global corporations, chambers of commerce are tethered intimately to the locations to which they pertain, offering an established channel of communication through which these businesses can potentially be mobilised into coordinated climate action (Verbovskii & Kosov, 2016).

Regarding mobilisation, chambers of commerce have an established rapport, confidence and familiarity among their network of businesses, which would perhaps generate representational equity of businesses in discussions and decision-making, offering greater buy-in for the commission objectives. Furthermore, they potentially tap into a wide set of SME organisations who may have larger (cumulative) carbon footprints than individual large corporations and are more likely to have local supply chain networks which contribute to the city's footprint.

EXTERNAL LENS: RESIDENTS' REPRESENTATIVES

Within the framework presented in Fig. 10.1, external motivators for businesses to engage in climate action consist of the economic and strategic benefits of a positive reputation derived from their climate engagement (Hoffman, 2016). In a place-based context, we therefore propose that one of the most important local external actors to businesses is the consumers in that city. As such, we suggest there is merit to the idea of involving local citizens in climate commissions, perhaps even ones for whom addressing climate change is not seen as a governance priority. As with the chamber of commerce example above, engagement with local businesses and their customers seems like an impactful strategy in the pursuit of buy-in for commission aims and objectives. Moreover, the merits of offering greater representation representing to a place's residents extend beyond the context of the private sector, in which this chapter is based, and can offer social and emotional perspectives which complement economic purposes for protecting cities from climate consequences.

Beyond achieving greater inclusiveness, the addition of residents' representation to climate commission compositions would ensure the entire spectrum of city actors, from governance bodies at the top to individual residents at the bottom, is accommodated in a unified city transition. Such an approach is also merited on the basis of some of the recognised failings of previous 'place-based' sustainability initiatives. For instance, the flagship policy of the 2010 UK Conservative Party, termed 'Big Society', fostered an inorganic sense of community due to the absence of local public input (Walker & Corbett, 2013). Similarly, the grassroot Transition Towns network, whilst generating social acceptance among residents, failed to make substantial transformations of localities as they failed to engage governing bodies to facilitate systematic change (Feola & Nunes, 2014). Indeed, the Leeds PCAN Climate Commission has noted that their own Citizen's Jury has offered constructive contributions through publicly announced recommendations, which have also furthered engagement, awareness and facilitated personal behaviour change among involved citizens (Leeds Climate Commission, 2019). Moreover, the Croydon Climate Crisis Commission also invites members of its local Citizen's Jury to sit on the commission as commissioners, offering the core PCAN commissions another opportunity to learn how to engage their own Citizen's Juries.

INTERNAL LENS: LOCAL TRADE UNION CHAPTERS

In contrast to external motivators, internal motivators are economic and strategic benefits of climate action that are realised within the business, namely employee productivity. Contemporary literature has identified that employees working for businesses that are proactive in their climate action derive a sense of heightened purpose, which has shown to positively correlate with greater productivity (Bocken & Geradts, 2020). As employees are responsible for the functional day-to-day operation of businesses, we identify the local workforce as a vital segment of the private sector, who warrant consideration for involvement in climate commissions. Whilst in the previous section we discussed the role of residents as consumers who offer a substantial collective agency in terms of pro (and anti-) environmental behaviour within the private sector, here we reflect on the extent to which more formal representation for employment-related issues offers another route to place-based buy-in for commissions.

By including local trade union chapters in their make-up, climate commissions would likely benefit from the established, vast network of employees across many industries within their local region that trades unions offer. Their cooperation could present the opportunity for established trades union networks to be used as a channel of communication to many residents of a city. Moreover, the pragmatic nature of trades unions, coupled with their knowledge of mobilising action, would possibly lend useful insight into climate commissions who aspire to muster a meaningful low-carbon transition. Furthermore, for an equitable transition to a low-carbon society to be achieved, the welfare of employees who work in high-emissions industries must be considered (Newell & Mulvaney, 2013). If a city hopes to achieve net-zero climate targets, businesses that are no longer viable in a green future must be reformed or made obsolete (Figueres & Rivett-Carnac, 2020). As such, trades unions could play a vital role in ensuring the fair transition of employees to new industries that are aligned with climate targets.

Legal Lens: Devolved Powers

In the context of place-based climate action, the main legal motivator for climate action that was identified in Fig. 10.1 was regulatory preparedness. The research found that businesses that have taken climate measures that go beyond expected climate regulation enjoy strategic and economic benefits to businesses that are continuously reacting to conform to new regulation (Bocken & Geradts, 2020). As actors must adapt in accordance with local regulation in order to be allowed to engage in business activities (Hoffman, 2016), the legal lens indicates that decision-making entities of local business regulation warrant a discernible segment in the private sector, as they play a significant role in steering it.

In a city context, local city councils are identified as senior governing bodies who often drive ambitious climate targets, such as the Edinburgh City Council (another place with a local PCAN Climate Commission) commitment to make the city climate neutral by 2030 (Edinburgh City

Council, 2020). Though it is acknowledged that city council representatives are often already climate commission members (indeed in Edinburgh the council was a founding partner of its local commission), the emphasis is predominantly on the council's own sustainability strategies. In addition to this, we suggest that commissions might prioritise legal experts who would offer climate commissions with a stronger grasp of the policy levers currently at the disposal of local authorities. As regularly advocated for by the Convention of Scottish Local Authorities (COSLA), there is an urgent need to devolve greater decision-maker power to local government on issues which impact sustainability and climate change (Davidson, 2019). There is an opportunity for climate commissions to join (or indeed bolster) the call for such changes.

RISK MITIGATION LENS: LAND AND REAL ESTATE OWNERS

Climate change risks in the context of the private sector can be broadly divided into two categories: *physical risks* and *market risks* (Cisar et al., 2011). *Physical risks* are the danger of physical damage to a business' assets due to climate change's consequences via extreme weather events. Alternatively, *market risks* emerge from the shifting business landscape prompted by climate change, which presents economic and competitive risks to businesses that do not transition effectively to align with a climate conscious market.

Regarding place, the *physical risks* are contextually nuanced—not every city will be subject to the same changing weather patterns—and the *market risks* vary from actor to actor within a place's business ecosystem (Figueres & Rivett-Carnac, 2020). In both cases, land and real estate owners (whose properties do not have the same footloose quality that capital benefits from) offer a potentially captive audience for serious conversations on addressing climate change. In addition to the risk of physical damage from extreme weather, COVID-19 has offered recent evidence of the way in which commercial spaces, office buildings and other urban land uses can lose significant value due to environmentally induced market shifts (Chernick et al., 2020). The risk mitigation lens, therefore, identifies landowners as an instrumental segment of the private sector. As there are many landowners within a city, we suggest that landowners with the largest portfolios are considered for involvement in climate commissions, as well as those that own land or businesses that will require significant reform to align with climate targets, and those at greatest imminent risks of climate change.

CONCLUSION

The purpose of this short chapter was to illustrate how a better understanding of the diversity actors in the private sector can pave the way for a more nuanced engagement with place-based stakeholder representation on climate commissions. By getting beyond an over-simplified narrative for the 'private sector' to play a role in combatting climate change, it is possible to think through how new and emerging city climate commissions can position themselves in ways that allow the diverse and varying resources at the disposal of situated private actors to be brought to bear on the climate challenge. The chapter drew on a comprehensive framework of private sector motivations to act on climate change in order to recommend stakeholders of a place who might offer important routes to the mobilisation of private sector resources in the pursuit of commission aspirations.

The financial lens showed the benefits that chambers of commerce would yield by offering a connection to businesses already embedded in 'place'; the external lens recommended a residents' representative to give a voice to consumers in the city; the internal lens recommended the participation of local chapters of trades unions, who may mobilise and represent the local workforce and ensure a just transition; the legal lens recommended incorporating intimate knowledge of the levers of power in governing the private sector currently at the disposal of local authorities; and finally, the risk mitigation lens proposed the addition of land and real estate owners given their vulnerabilities to a changing climate.

Though this book chapter has outlined various actors that we believe would offer valuable private sector representation on commissions, it does not aim to suggest what roles these new commissioners might assume. Rather, we wish to demonstrate that the setting-up of a city-based climate commission, as an attempt to institutionalise place-based experimental governance, warrants nuanced consideration of who it is that represents the private sector in ways that can best mobilise its diverse resources at the local scale. There is a need for close engagement with the unique setting of any given city, particularly with regard to the relationship (or lack of it) between private businesses and the political, institutional and cultural nature of the places in which they are located. The introduction of some of the diverse actors presented in this chapter (actors who might not be the most obvious starting point for taking action on climate change) could allow a more intimate and practical engagement with the private sector, and ultimately generate meaningful climate action at the city-scale.

REFERENCES

Averchenkova, A., Frankhauser, S., & Finnegan, J. J. (2020). The impact of strategic climate legislation: Evidence from expert interviews on the UK Climate Change Act. *Climate Policy*, 1–13.

Bocken, N. M., & Geradts, T. H. (2020). Barriers and drivers to sustainable business model innovation: Organization design and dynamic capabilities. *Long Range Planning*, 53(4), 101950.

Chernick, H., Copeland, D., & Reshovsky, A. (2020). The fiscal effects of the COVID-19 pandemic on cities: An initial assessment. *National Tax Journal*, 73(3), 699–732.

Cisar, J., et al. (2011). Physical and economic consequences of climate change in Europe. *Proceedings of the National Academy of Sciences of the United States of America*, 108(7), 2678–2683.

Davidson, J. (2019). *COSLA president:Local government has 'not benefited from devolution'* [Online]. Available at: https://www.holyrood.com/news/view,cosla-president-local-government-has-not-benefited-from-devolution_1 4538.htm. [Accessed 14 January 2021].

Edinburgh Chamber of Commerce. (2020). *What we do* [Online]. Available at: https://www.edinburghchamber.co.uk/about-us/what-we-do/. Accessed 13 January 2021.

Edinburgh City Council. (2020). *Our climate target: Net-zero by 2030* [Online]. Available at: https://www.edinburgh.gov.uk/climate-2/climate-target-net-zero-2030/1#:~:text=Read%20article%20about%20Edinburgh's%20commitm ent,city%2C%20put%20into%20the%20air. Accessed 14 January 2020.

Evans, S., et al. (2017). Business model innovation for sustainability: Towards a unified perspective for creation of sustainable business models. *Business Strategy and the Environment*, 26(5), 597–608.

Feola, G., & Nunes, R. (2014). Success and failure of grassroot innovations for addressing climate change: The case of the transition movement. *Global Environmental Change*, 24, 232–250.

Figueres, C., & Rivett-Carnac, T. (2020). *The future we choose: Surviving the climate crisis* (1st ed.). Knopf.

Hawken, P. (2017). *Drawdown: The most comprehensive plan ever proposed to reverse global warming* (2nd ed.). Penguin.

Hoffman, A. J. (2016). Communicating about climate change with corporate leaders and stakeholders. In M. Nisbet (Ed.), *The encyclopedia of climate change communication*. Oxford University Press.

Leeds Climate Commission. (2019). *Leeds climate change citizen's jury* [Online]. Available at: https://www.leedsclimate.org.uk/leeds-climate-change-citizens-jury. Accessed 8 March 2021.

Newell, P., & Mulvaney, D. (2013). The political economy of the 'just transition'. *The Geographical Journal, 170*(2), 132–140.

Place-Based Climate Action Network (PCAN). (2020). *Driving climate action in UK cities and communities* [Online]. Available at: https://www.pcancities.org.uk/. Accessed 13 January 2021.

Schaltegger, S., Lüdeke-Freund, F., & Hansen, E. G. (2012). Business cases for sustainability: The role of business model innovation for corporate sustainability. *International Journal of Innovation and Sustainable Development, 6*(2), 95–119.

Stern, N. (2007). *The economics of climate change—The Stern review.* Cambridge University Press.

Torraco, R. J. (2005). Writing integrative literature reviews: Guidelines and examples. *Human Resource Development Review, 4*(5), 356–367.

Verbovskii, V., & Kosov, V. (2016). Five reasons to join local Chambers of Commerce and Industry. *Journal of Economics and Social Sciences, 8.*

Walker, A., & Corbett, S. (2013). The 'big society', neoliberalism and the rediscovery of the 'social' in Britain. *SPERI.*

Open Access This chapter is licensed under the terms of the Creative Commons Attribution 4.0 International License (http://creativecommons.org/licenses/by/4.0/), which permits use, sharing, adaptation, distribution and reproduction in any medium or format, as long as you give appropriate credit to the original author(s) and the source, provide a link to the Creative Commons license and indicate if changes were made.

The images or other third party material in this chapter are included in the chapter's Creative Commons license, unless indicated otherwise in a credit line to the material. If material is not included in the chapter's Creative Commons license and your intended use is not permitted by statutory regulation or exceeds the permitted use, you will need to obtain permission directly from the copyright holder.

CHAPTER 11

Citizens' Assemblies and Juries on Climate Change: Lessons from Their Use in Practice

Rebecca Wells

Highlights Citizen assemblies and juries (CAJs) must meet generally accepted standards and be citizen-led to genuinely and credibly engage citizens. Agreed implementation and follow-up procedures should be established to ensure CAJs legitimately inform policymaking. CAJs are not a panacea to public participation on climate change and much more needs to be done beyond them.

Keywords Citizen assembly · Citizen jury · Climate change · Democratic deliberative process · Citizen engagement

INTRODUCTION

Globally, 64% of people believe there is a 'climate emergency' (UNDP, 2021). Since Bristol declared a 'climate emergency' in November 2018, over 300 councils and the UK Parliament have followed suit (Declare A

R. Wells (✉)
London School of Economics, London, UK

© The Author(s) 2022
C. Howarth et al. (eds.), *Addressing the Climate Crisis*,
https://doi.org/10.1007/978-3-030-79739-3_11

Climate Emergency, 2020). In response to this climate emergency narra-tive, many national and local governments have turned to deliberative democratic processes such as citizen assemblies and juries (CAJs) as tools to gauge public opinion to inform their responses to the climate crisis (Mellier-Wilson & Toy, 2020). Deliberative processes remain relatively marginal, but CAJs on climate change specifically have recently emerged in the UK and abroad (Devaney et al., 2020). CAJs can significantly contribute to engaging more deeply with the public on the climate crisis and creating more inclusive, citizen-driven policymaking and are widely supported in academia and by activist groups such as Extinction Rebel-lion (XR) (Devaney et al., 2020). However, it is important to critically analyse how CAJs are driving change in practice. To enable learning and improve future CAJs, this chapter identifies how they could be improved by assessing their quality and the impact CAJs have had on policymaking.

Academic and grey literature on the use of CAJs as a method to increase public engagement in climate policymaking was reviewed, followed by a comparative analysis of the reports produced by completed CAJs and the responses of governing bodies in the UK as well at the national level in the UK, Ireland and France. Areas where their use in practice differed from each other and deviated from the literature were identified in order to determine how future processes could be improved to enhance their ability to increase public engagement on climate change.

ARE CLIMATE ASSEMBLIES AND JURIES USEFUL IN TACKLING CLIMATE CHANGE?

Tools of deliberative democracy are methods to engage with the public to create a structured dialogue between citizens, experts and politi-cians in order to help politicians understand public views on different policy approaches, creating more informed political decision-making and thereby increasing the democratic legitimacy of policies (Howarth et al., 2020; Willis, 2020). National governments usually make top-down climate policy decisions with little input from the public and lack a clear sense of the wider public's mandate for climate action. Yet, few attempts have been made to engage the public in the need for, and benefits of, tran-sitioning to a net-zero carbon society (Willis, 2019). Broader and more direct public participation in climate policymaking has been widely advo-cated as a way to increase the legitimacy and quality of policy decisions, and a failure to do so risks public backlash, such as the 'gilets jaunes' protests which emerged in France (Dietz & Stern, 2008; Kythreotis et al., 2019).

CAJs are deliberative tools to engage with citizens and supplement representative democracy by bringing informed citizens' perspectives into the decision-making process (Smith & Wales, 2000). The processes are similar but citizen assemblies usually include a representative sample of 50–160 and citizen juries include 12–30 people in the target population, a group small enough to be genuinely deliberative but large enough to be representative (Bryant, 2019; Roberts & Escobar, 2015; Goodin & Dryzek, 2006). An independent oversight panel consisting of key stakeholders oversees the process (Wakeford et al., 2015). Participants receive and cross-examine expert information on a particular issue and deliberate with each other, discussing different perspectives and trade-offs to propose a series of informed and considered recommendations to deal with that issue (Goodin & Dryzek, 2006; Roberts & Escobar, 2015; Smith & Wales, 2000).

CAJs can potentially provide better insight into public opinions on climate change that have been reached in a fair and informed way, allowing citizens to test and discuss a range of approaches to climate action whilst facilitating public support for tough policy decisions by including the concerns and ideas of citizens in policymaking (Willis, 2018; Devaney et al., 2020; Bryant, 2019). However, their use in practice must be analysed to determine the extent to which they achieve and demonstrate these deliberative benefits.

IMPACT ON POLICYMAKING

CAJs on climate change tend to generate very ambitious recommendations (Willis, 2020). For example, Wilson and Mellier (2020) claim that both the UK and French Climate Assemblies generated far more ambitious policies than politicians have ever proposed. Bryant and Stone (2020) argue that CAJs' biggest impact is to create a strong political platform for action by providing elected representatives with a public mandate on climate change. Often CAJs are followed by increased climate action, such as in Oxford where the council announced over £1 million additional funding and £18 million of capital investment to address climate change along with a range of commitments in response to their Citizens Assembly (Oxford City Council, 2019).

However, the recommendations produced in CAJs often have an advisory role and compete with advice from other groups, making their impact on policymaking difficult to identify (Bryant & Hall, 2017; Flinders et al., 2015). In most cases, commissioning bodies respond to the recommendations in reference to current policies and claim they will inform an upcoming climate plan. For example, the recommendations produced by the Brent Climate Change Citizens' Assembly (November–December 2019) seem to have had a strong influence over the 2021–2030 Brent Climate Emergency Strategy, which refers to the recommendations throughout (Brent Council, 2020). However, this link is not always obvious. As another example, after the UK Climate Assembly (January–May 2020), the convening parliamentary committees, the government and the Climate Change Committee committed to take the recommendations on board, although it is not clear how (Bouyé, 2020). Dicker (2020) considers the limitations of the process and argues that the UK Climate Assembly could have been improved by having a direct link to legislative, policy and funding decisions as it had no mandate from, or direct link, to government.

The French Climate Assembly (October 2019–June 2020) could have had a large impact as President Macron gave it the power to generate policies that could be enacted either through a national referendum, parliamentary vote or directly through executive orders (Wilson & Mellier, 2020; O'Grady, 2021). However, almost a year after the Assembly, its members rated the French government's proposed climate and resilience law 3.3 out of 10 for reflecting their recommendations, which suggest that its impact has been far from that which was promised (Climate Home News, 2021). However, recommendations arguably should not be directly implemented, as CAJs are not authorised to govern through democratic processes such as elections. CAJs' lack of democratic validity suggests that they should act as an advisory body complemented by further expertise and evidence-based input (Devaney et al., 2020; O'Grady, 2021). For example, after the Irish Citizens' Assembly, an all-party parliamentary committee was established to respond to the recommendations on climate change. This committee published a report largely endorsing and further developing the Assembly's recommendations, which had a significant role in advising and shaping the development of the Irish government's 2019 Climate Action Plan to Tackle the Climate Breakdown (Devaney et al., 2020). Thus, there is a large variation in how CAJs are integrated into policymaking in practice.

There is a need for agreed follow-up and implementation procedures for the recommendations produced by CAJs including a guaranteed response from the commissioning body (Devaney et al., 2020). Ensuring that recommendations from a CAJ are incorporated into the policymaking processes in an appropriate and transparent manner is vital to ensure they are seen as legitimately integrating citizens' views into policymaking (Devaney et al., 2020). Nevertheless, CAJs provide a strong mandate and momentum for climate action which allows policymakers to introduce more drastic policies, as often seen in practice where stronger climate policies are announced following them (O'Grady, 2021).

ARE THESE TOKENISTIC PROCESSES?

There is a risk that CAJs are being used as a tokenistic exercise, enabling governing bodies to claim that public opinion has been considered, rather than building a genuine dialogue between them and the public. One key indication of this is that some processes being labelled CAJs do not meet the generally accepted standards for them. For example, the Deputy City Mayor of Leicester (Clarke, 2020) admitted that their Climate Assembly did not qualify as a citizens' assembly as its method of recruitment 'didn't match that of a jury or citizens' assembly' and the process was only one day long (Clarke, 2020). Similarly, the Camden Citizens' Assembly on the Climate Crisis only totalled 12 hours (Cain & Moore, 2019). Neither of these cases meet generally accepted standards for CAJs which should randomly select participants from the population and be at least 20 hours in length to allow proper learning and deliberation to occur (Cain & Moore, 2019; Mellier-Wilson & Toy, 2020). This suggests that some engagement processes are wrongly labelled CAJs because they are currently a trendy engagement method and seen as good politics. For CAJs to truly realise their potential as deliberative processes and be seen as legitimate long-term forms of public engagement on climate change, they must be done rigorously and meet generally accepted standards (Bryant, 2019; O'Grady, 2021).

Those processes being run with a more consultative structure where participants prioritise a pre-prepared list of policy options versus those which allow participants to come up with their own recommendations also run the risk of being used as tokenistic exercises (Bryant &

Stone, 2020). For example, the French Climate Assembly was citizen-led as it was a political chamber where citizens came up with legislative proposals, which could be directly passed into law (Wilson & Mellier, 2020). In contrast, the recommendations produced by the UK Climate Assembly were based on predetermined policy options meaning that citizens were not able to shape the agenda, process or come up with their own measures, instead considering those already drafted by government (Wilson & Mellier, 2020). A more consultative structure may be little more than a short-term consultation for interested parties to give the appearance of public legitimacy to political decisions that have already been made (Wakeford et al., 2015). Therefore, future processes should aim to be citizen-led to allow public concerns to be truly considered in policymaking.

WIDER PUBLIC ENGAGEMENT

CAJs have the potential to ignite wider public debates on climate change. Going back to the example of the French Climate Assembly, it generated a genuine national debate. 70% of people in France knew of the Assembly, and of those, 64% considered its work useful to fight against climate change (Resau Action Climat France, 2020). Thus, the Assembly generated a powerful mandate for change but also a movement of people who engaged with the Assembly itself (Wilson & Mellier, 2020). CAJs run in the UK have largely failed to ignite a wider public debate, often due to budget limitations and integrated planning but also because CAJs are rarely seen as tools which can start a wider public dialogue (Bryant & Stone, 2020). This is a missed opportunity as CAJs should aim to generate a public debate to increase momentum and hold governing bodies accountable for the recommendations (Wilson & Mellier, 2020).

However, CAJs are not a panacea for solving issues with public participation and climate policymaking (Smith & Wales, 2000; Devaney et al., 2020; Flinders et al., 2015). CAJs only represent one form of public engagement and deliberation on climate change, and there are a variety of other communications, education and engagement initiatives available (Devaney et al., 2020). Additionally, public engagement with climate change is required beyond the formal process of CAJs so that people better understand and can help shape low-carbon transformations (Capstick et al., 2020; Devaney et al., 2020). This is demonstrated by the fact that the recommendations produced by almost all CAJs request more education and engagement with citizens on climate change. For example, 8 out of the 25 recommendations produced by the Lancaster district Climate Change People's Jury revolved around improving communications, education and council leadership on climate change (Shared Future, 2020). Therefore, whilst CAJs are a positive step towards increasing public engagement on climate change, much more needs to be done to engage with citizens on this issue. CAJs should be used alongside other tools to engage the public and enable them to play a role in climate change policymaking.

CONCLUSION

Overall, this chapter highlights how the use of CAJs in practice must be critically assessed to allow future CAJs to be improved and have maximum impact on climate action in practice. There is a need for agreed follow-up and implementation procedures to increase transparency in how CAJs create more citizen-centred policymaking and prevent their use becoming tokenistic. The structure of CAJs varies in practice, impacting the extent to which they truly incorporate citizens' views into the construction of climate policies. Thus, CAJs must be designed carefully to enable their potential benefits to be realised. Furthermore, processes which claim to be CAJs on climate change should meet generally accepted standards to ensure that they represent rigorous deliberative processes and are not tokenistic exercises being used to give the illusion that public opinion has been taken into account in policymaking.

The limitations of CAJs must also be considered when they are being designed and used. For example, CAJs only include a small proportion of the target population, so their representativeness is not a given and their recruitment processes must be robust if their outcomes are to be truly representative.

Nevertheless, CAJs provide an opportunity to gather views on climate change of an informed and representative group of the target population. CAJs can also engage the wider public in climate change debates, an opportunity that future CAJs should seize in order to maximise their impact. However, CAJs only represent one form of citizen engagement and are not a panacea to tackling issues around public engagement on climate change. Thus, whilst their expanding use can increase public engagement on climate change, they cannot be the only mechanism to do so.

REFERENCES

Bouyé, M., 2020. *Early lessons from France and the UK on the roles of climate citizens' assemblies and legislators to enhance climate action* [Online]. Available at: https://www.wfd.org/2020/10/13/early-lessons-from-france-and-the-uk-on-the-roles-of-climate-citizens-assemblies-and-legislators-to-enhance-climate-action/. Accessed 28 December 2020.

Brent Council. (2020). *Brent Climate Emergency Strategy 2021–2030*. Brent Council.

Bryant, P. & Hall, J. (2017). *Citizens Jury Literature Review*. S.l.: Shared Future.

Bryant, P. (2019). *Citizens assemblies, citizens' juries and climate change* [Online]. Available at: https://sharedfuturecic.org.uk/citizens-assemblies-citizens-juries-and-climate-change/. Accessed 2 January 2021.

Bryant, P., & Stone, L. (2020). *Climate assemblies and juries: A people powered response to the climate emergency: A guide for local authorities and other bodies*. Shared Future.

Cain, L., & Moore, G. (2019). *Evaluation of Camden Council's Citizens' Assembly on the climate crisis*. UCL.

Capstick, S., et al. (2020). *Climate change citizens' assemblies* (CAST Briefing Paper 03). The Centre for Climate Change and Social Transformations (CAST).

Clarke, A. (2020). *Making a space for deliberative democracy at Leicester City Council: The Leicester Climate Assembly* [Online]. Available at: https://www.thersa.org/blog/2020/02/making-a-space-for-deliberative-democracy-at-leicester-city-council-the-leicester-climate-assembly#:~:text=What%20mate rialised%20was%20the%20Leicester,'Leicester%20in%20a%20room. Accessed 22 December 2020.

Climate Home News. (2021). *French climate bill set for rocky ride after citizens' assembly slams weak ambition* [Online]. Available at: https://www.climatech angenews.com/2021/03/03/french-climate-bill-set-rocky-ride-citizens-ass embly-slams-weak-ambition/. Accessed 8 March 2021.

Declare A Climate Emergency. (2020). *List of councils who have declared a climate emergency* [Online]. Available at: https://www.climateemergency.uk/blog/list-of-councils/. Accessed 20 December 2020.

Devaney, L., Torney, D., Brereton, P., & Coleman, M. (2020). *Deepening public engagement on climate change: Lessons from the citizens' assembly*. Environmental Protection Agency.

Dicker, S. (2020). *Where next for the UK Climate Assembly?* [Online]. Available at: https://www.lse.ac.uk/granthaminstitute/news/where-next-for-the-uk-cli mate-assembly/. Accessed 19 December 2020.

Dietz, T., & Stern, P. C. (2008). *Public participation in environmental assessment and decision making* [Online]. Available at: https://doi.org/10.3389/fenvs. 2019.00010. Accessed 1 May 2020.

Flinders, M., et al. (2015). *Democracy matters: Lessons from the 2015 citizens' assemblies on English devolution*. Democracy Matters.

Goodin, R. E., & Dryzek, J. S. (2006). Deliberative impacts: The macro-political uptake of mini-publics. *Politics & Society, 34*(2), 219–244.

Howarth, C., et al. (2020). Building a social mandate for climate action: Lessons from COVID-19. *Environmental and Resource Economics: Special Issue 'Environmental Economics in the Shadow of Coronavirus'*, Volume in Press.

Kythreotis, A. P., et al. (2019). Citizen social science for more integrative and effective climate action: A science-policy perspective. *Frontier Environmental Science*.

Mellier-Wilson, C., & Toy, S. (2020). *UK climate change citizens' assemblies & citizens' juries* [Online]. Available at: https://www.involve.org.uk/resources/case-studies/uk-climate-change-citizens-assemblies-citizens-juries. Accessed 2 January 2021.

Mellier, C., & Wilson, R. (2020). *Getting climate citizens' assemblies right* [Online]. Available at: https://carnegieeurope.eu/2020/11/05/getting-climate-citizens-assemblies-right-pub-83133. Accessed 10 December 2020.

O'Grady, C. (2021). Power to the people: Nations are turning to citizen assemblies to weigh up climate policies. *Science, 370*(6516), 518–521.

Oxford City Council. (2019). *City Council responds to Oxford citizens' assembly on climate change and outlines 19 million pound climate emergency budget* [Online]. Available at: https://www.oxford.gov.uk/news/article/1275/city_council_responds_to_oxford_citizens_assembly_on_climate_change_and_outlines_19m_climate_emergency_budget#:~:text=Oxford%20City%20Council%20has%20responded,Zero%20Carbon%20Council%20and%20city. Accessed 20 October 2020.

Reseau Action Climat France. (2020). *Survey: Gauls not so refractory to climate action* [Online]. Available at: https://reseauactionclimat.org/sondage-des-gaulois-pas-si-refractaires-a-laction-climatique/. Accessed 7 March 2021.

Roberts, J., & Escobar, O. (2015). *Involving communities in deliberation: A study of three citizens' juries on onshore wind farms in Scotland*. ClimateXChange.

Shared Future. (2020). *Lancaster district climate change people's jury recommendations*. Shared Future; Lancaster City Council.

Smith, G., & Wales, C. (2000). Citizens juries' and deliberative democracy. *Political Studies, 48*, 51–65.

UNDP. (2021). *The peoples' climate vote* [Online]. Available at https://www.undp.org/content/undp/en/home/librarypage/climate-and-disaster-resilience-/The-Peoples-Climate-Vote-Results.html. Accessed 30 March 2021.

Wakeford, T., Walcon, E., & Pimbert, M. (2015). Refashioning citizens' juries: Participatory democracy in action. In H. Bradbury-Huang (Ed.), *The Sage handbook of action research* (pp. 230–247). Sage.

Willis, R. (2018). *Building the political mandate for climate action*. London: Green Alliance.

Willis, R. (2019). *Citizens' assemblies and citizens' juries: What happens next?* [Online]. Available at: https://www.rebeccawillis.co.uk/citizens-assemblies-and-citizens-juries-what-happens-next/. Accessed 15 June 2020.

Willis, R. (2020). *Too hot to handle? The democratic challenge of climate change.* Bristol University Press.

Open Access This chapter is licensed under the terms of the Creative Commons Attribution 4.0 International License (http://creativecommons.org/licenses/by/4.0/), which permits use, sharing, adaptation, distribution and reproduction in any medium or format, as long as you give appropriate credit to the original author(s) and the source, provide a link to the Creative Commons license and indicate if changes were made.

The images or other third party material in this chapter are included in the chapter's Creative Commons license, unless indicated otherwise in a credit line to the material. If material is not included in the chapter's Creative Commons license and your intended use is not permitted by statutory regulation or exceeds the permitted use, you will need to obtain permission directly from the copyright holder.

CHAPTER 12

Universities as Living Labs for Climate Praxis

*Zoe P. Robinson, Philip Catney, Philippa Calver,
and Adam Peacock*

Highlights University living lab success relies on careful navigation of complex relationships between different actors. Maximising change through living labs requires educational objectives and learning processes embedded in governance.

Keywords Climate change · Living labs · Universities · Campuses · Climate praxis

Z. P. Robinson (✉) · A. Peacock
Institute for Sustainable Futures/School of Geography, Geology and the Environment, Keele University, Keele, UK
e-mail: z.p.robinson@keele.ac.uk

P. Catney
School of Social, Political and Global Studies, Keele University, Keele, UK

P. Calver
Tyndall Centre for Climate Change Research, University of Manchester, Manchester, UK

© The Author(s) 2022
C. Howarth et al. (eds.), *Addressing the Climate Crisis*,
https://doi.org/10.1007/978-3-030-79739-3_12

INTRODUCTION

Climate imperatives have led to an increased interest in the use of 'living labs' as places of experimentation and innovation for climate-related solutions. The term living lab in the context of this paper is used to describe the combination of a place (e.g. a university or city) and a research and built environment management approach where innovations are trialled in the 'real world' for the purpose of both making sustainable improvements and generating learning about the effectiveness of the solution and implementation process. Living Labs can occur at different scales, but are typically viewed as geographically bounded and involving intentional interventions and feedback loops to facilitate adaptive learning, and experimental forms of collaborative governance between diverse stakeholders (Evans & Karvonen, 2014; Evans et al., 2015). Alongside the development of urban living labs, developed in partnership between government and public and private property owners (see Evans & Karvonen, 2014), university campuses provide additional venues for the development of living labs as places of climate praxis. Both settings share some characteristics; however, each has their own opportunities and challenges.

Universities can impact climate praxis in many ways, one example is that, as sizable organisations, they have significant energy demands. An increasing imperative to reduce carbon emissions from their estates is reflected in the numerous university declarations of a climate emergency and ambitious net-zero targets. Universities have the potential to catalyse wider-scale changes in climate praxis through education, research and business engagement activities. Additionally, with a growing emphasis on partnerships with industry and government, universities are increasingly playing an important part in their wider regions as 'anchor institutions', providing leadership and support on issues such as economic development, health and environmental matters (Birch et al., 2013). These factors demonstrate that university campuses are 'privileged space[s] of innovation' (Evans & Karvonen, 2014, p. 415), offering the potential to trial new governance approaches and technologies for climate change mitigation in ways that may be difficult to undertake in other public settings.

It is critical to recognise that whilst all universities can contribute to climate praxis, not all are equally placed to perform as living labs. Factors that contribute to universities as propitious places for exploring climate solutions include independent management of their utility networks, control over a multi-use built environment (including retail, catering, leisure, conferencing, offices, residential and laboratory facilities) and a significant and diverse community of staff and students who may live and work within the university campus ecosystem. Often this campus ecosystem represents the scale and complexity of a small town, in turn securing their potential as living labs (Colding & Barthel, 2017).

Even where this combination of useful characteristics is present, significant challenges remain. For example, the complexity of the university ecosystem, one of the qualities that make university living labs so attractive in the first place, needs to be considered explicitly as part of project design and implementation. To this end, this chapter draws on reflections from two university-based, campus-scale sustainable energy-transition 'living lab' projects to explore two key areas of challenges and opportunities that require consideration if we are to maximise the potential of universities as places of climate praxis.

After outlining the projects and setting for the living lab, our first reflection explores the experiences of those in the living lab and how these may be mediated by their relationships with the university and the complexity of their interlinked private and public spheres. In addition, we highlight the need for project implementers to be sensitive to the position and views of the living lab 'users'. The second reflection explores how to harness university-based living lab projects to address specific educational opportunities within a university living lab, and enhance climate change and energy literacy, helping to prepare society for sustainable transitions.

The Living Lab Projects

The reflections of the two energy-transition projects based at one university are drawn from the different positionalities and roles of the authors in relation to the projects. These include formal academic representation in project governance, data collection, membership of the staff, student, and resident community, and roles pertaining to sustainability governance in the university. The reflections are also influenced by interview data collected from research carried out in relation to both projects, comprising 27 interviews with a range of project stakeholders across both projects. The university is not named for anonymity purposes.

Whilst both projects relate to climate praxis through greenhouse gas reduction in the university's energy systems, the projects' foci differed.

The first project aimed to demonstrate the safe, efficient and non-disruptive distribution of blended hydrogen in the gas network. The second project focused on the development of a smart energy network management system linked to a significant increase in onsite renewable electricity generation. Both were multi-million-pound projects supported by public funding. The hydrogen project was led by one of the UK's gas distribution network operators, alongside the university and industrial partners. The project utilised the university's private gas network as the first trial stage prior to a public site trial. In contrast, the smart energy network project was led by the university, with a major engineering multinational corporation as the key design and delivery partner. This project aimed to make the campus a research and development facility, creating an at-scale living lab where smart energy strategies and technologies could be researched, developed and tested in a real-world environment, whilst also delivering against ambitious onsite carbon reduction targets.

The university hosting these projects is a semi-rural campus university with a student population of 10,000 and a large campus estate of 600 acres. As well as accommodating over 3,000 students on campus (largely in on-campus halls of residence), there are over 100 properties on campus for staff (and former staff) residents. These properties range from flats to detached houses and have a mix of owner-occupied and rental properties where the university acts as the landlord. A proportion of residences on campus are second homes for staff who have permanent residences at a distance from the university, and only spend a portion of a week or the year living on the campus, whilst other campus residences are permanent homes for staff and their families. An array of catering, leisure and retail amenities also exists. These aspects, alongside its private utility networks, make it a particularly attractive site as a living lab for at-scale climate praxis innovations.

Understanding the Experience of Living in a Living Lab

Key to the concept of a living lab is that there are 'users' interacting with its technological and governance systems as part of their normal routine. In the context of our case studies, the users range from the staff and student residents and campus users to the estates-based staff with responsibility for the operation and services of the built environment.

Reflecting on interviews with the living lab residents highlights the diverse views that can exist about being part of such projects. Attitudes towards and prior experiences with the university itself appear to strongly influence residents' perceptions of projects and their willingness to actively engage. Whilst our interviews demonstrated that some residents may feel very positive about the projects and pride in their and the university's involvement, other residents showed some dissatisfaction.

One issue that was raised was how households are recruited to be part of the living lab and the limited opportunity to 'opt out' with some residents believing the university was making proprietary decisions over the residents' private spaces. In a university living lab where many of the 'users' are academics engaging regularly and explicitly with issues of research ethics, sensitivities on the issue of consent may be particularly heightened. This needs careful consideration and management by project implementers particularly if there are longer-term aspirations to include the whole campus environment as a living lab. These issues reflect wider debates about the role of informed consent as an integral element of justice within the transition to a new energy future (Sovacool & Dworkin, 2015). Whilst these issues are relevant for all living labs, there are important nuances in a university setting due to the complex relationships between project implementers and living lab 'users'.

In addition, privacy concerns for some influenced project implementation. Specifically, some campus households were unhappy about the request for smart meter installation, mirroring privacy concerns in the general population (see, for example, McKenna et al., 2012). We found the desire for privacy potentially compounded by increased sensitivities relating to the complex relationship between resident and university employer, and interestingly heightened in times of university-wide work-related disputes. These reflections underscore the need for project implementers to be sensitive to the ways in which users' perception of a technology is potentially entangled with their social identities and their contexts in complex and dynamic ways (Callon & Rabeharisoa, 2003), and how this plays out in a university setting.

Whilst residents' prior experiences of, and assumptions about, the university play out in their perceptions of being involved in a university living lab, project implementers' assumptions about, and previous experiences of, residents and other users may also influence how residents and other users are incorporated into and engaged within projects. Our reflections demonstrate that residents often viewed themselves as active stakeholders in the projects with an active interest in and expectations of being kept informed about the projects. In contrast, project implementers wanted to reduce disturbing residents by limiting communication when no direct resident input was seen as necessary, leading to dissatisfaction for some residents.

Our reflections on interviews with diverse living lab stakeholders and our own positionalities highlight the need for project implementers to acknowledge the important role of users within a living lab, recognising these as key, often very engaged stakeholders, and hence the need to ensure that their voices are heard. Effectively embedding users at the centre of living labs requires community engagement expertise and necessitates a model of governance adaptable to the needs of users, which are not necessarily automatically part of standard university estate project implementation. Rather than seeing engagement as a single-stage or outcome in the project delivery, a greater focus on user-centred governance can contribute to project success by delivering instrumental benefits (increasing participant engagement and project support), substantive benefits (where greater communication leads to better-informed decision-making) and normative benefits (where a 'just' process is developed). Investing time and money into ongoing engagement with user communities should be prioritised at the highest level in the governance of all living lab climate praxis projects.

Maximising Learning and Preparing Society for the Sustainable Energy Transition

Universities are places of learning. Yet learning takes place not just through the formal curriculum, but through informal learning opportunities, including activities outside the classroom and from the campus environment itself. University living labs therefore may offer the opportunity to capitalise on this learning mission, to enhance climate praxis and energy-literacy learning among all its stakeholders.

Increasing citizens' 'energy literacy', the understanding of energy and its role in society has two important functions. First, it allows informed citizen engagement with energy decision-making, which many authors believe will subsequently increase individual support for investment in a low carbon pathway (DeWaters et al., 2013). Second, it enables decision-making over personal energy practices to be informed by energy realities (Hogan et al., 2019; Martins et al., 2020). We are living through an energy paradigm shift, characterised by the decarbonisation, decentralisation, digitisation and democratisation of our energy systems (Becker & Naumann, 2017). However, little of the current transitions towards this shift is visible to the wider public. This 'invisibility' of parts of the energy system could in part explain low levels of energy literacy among citizens (Cotton et al., 2016).

Research on the energy literacy of students has highlighted both the patchy energy knowledge of students and the potential for sustainability and by extension energy learning, from the campus environment (Cotton et al., 2015). A further potential benefit of user engagement around specific campus-based energy projects is the potential to catalyse wider engagement with pro-environmental behaviours, due to increased environmental consciousness-raising. Therefore, university living lab climate praxis projects provide the potential to increase the energy literacy of campus energy users, increase users' energy-transition readiness as well as potentially promote wider pro-environmental behaviours.

Our reflections on the communication and engagement of the climate praxis projects discussed here are that they were largely focused on individuals affected by the projects in their private spaces, with more limited engagement with energy users within the university's public spaces, such as building managers, building users and students using university facilities and in halls of residence. By omitting a deeper level of engagement with these wider, diverse 'public' energy user audiences, potentially important opportunities to enhance wider engagement with energy transitions and climate praxis are missed. To ensure such learning opportunities are not missed, effective wider stakeholder engagement with climate praxis projects is required across the diverse energy users on a university campus, and educational goals need to be embedded in project governance from the start. This seems particularly important for a university living lab within the context of a university's core educational mission and the opportunities to integrate students into living lab activities. However, considering the potential for further learning of all stakeholders in any living lab should also be considered explicitly.

A final area of reflection concerning maximising learning is how university project teams learn from their own practices and experiences, and how the characteristic of flexible, adaptive governance required by living labs is ensured (Evans et al., 2015). Our path to a sustainable future is not predetermined and requires reflexive learning and governance structured around multiple stakeholders and co-creation (McCrory et al., 2020), as well as different disciplinary perspectives to maximise learning and test assumptions. Therefore, project teams should adopt the principles of a 'learning organisation' with diverse project stakeholders working together to improve capacities and transform practice (Senge, 2006). Organisational learning could be achieved by embedding research and evaluation of the 'living lab' process and user experience into project governance. However, project teams should go further and develop a genuinely cross-university community of practice (Wenger, 1999), drawing together different areas of expertise to help overcome the barriers to effective joint-action and create more legitimacy for the living lab and the broader goals of sustainability. A community of practice should not be exclusive to those working within the project teams, but should include more interaction with the users and their communities in which living labs are situated to develop a more nuanced understanding of the context for living lab interventions but also to enable greater social learning.

CONCLUSION

Universities are important spaces as living labs for exploring sustainable solutions due to the mixed-use built environment, private infrastructure and potential to link with the research, education and business engagement missions of the university. However, there is also a need to recognise the complexities, sensitivities, challenges and differences that may be particular to university settings as living labs.

To maximise the potential of universities as living labs for climate praxis requires:

1. Sensitivity to the lived experience of those within the living lab, with careful consideration in climate praxis projects of the role of informed consent in participation, the implications of overlap between the public and private spheres within the university ecosystem and attention to the complex, multifaceted relationships that exist between members of the university community.
2. Effective communication with all project stakeholders and the wider university community, drawing in expertise in community engagement and communication, ensuring sufficient frequency and depth to respect the role of stakeholders as energy users in the living lab, as well as enhance wider learning and energy-transition readiness, and potentially catalyse further pro-environmental behaviour.

3. Project design and governance that allow reflexive learning of climate praxis within the university and encompass the wider educational and research missions and interdisciplinarity inherent in a university, and that embed mechanisms for learning from user experiences and the 'living lab' process itself.

University living labs can support carbon reduction on the estate itself and share learning to be utilised for other university campuses; they can also provide distinct testbed environments for climate praxis interventions that can be utilised outside of the university environment. However, care needs to be taken to consider the transferability of learning between the university context and wider environment. Although many living lab projects on universities may focus on a core goal of emissions reductions from the estate, universities stand apart as places of learning and research. To maximise the real potential of universities as living labs, the design and governance of climate praxis projects must not just focus on the estate, but the learning to be gained through their education and research missions and the potential to disseminate such learning.

References

Becker, S., & Naumann, M. (2017). Energy democracy: Mapping the debate on energy alternatives. *Geography Compass, 11*(8), e12321.

Birch, E., Perry, D. C., & Taylor, H. L., Jr. (2013). Universities as anchor institutions. *Journal of Higher Education Outreach and Engagement, 17*(3), 7–15.

Callon, M., & Rabeharisoa, V. (2003). Research "in the wild" and the shaping of new social identities. *Technology in Society, 25*(2), 193–204.

Colding, J., & Barthel, S. (2017). The role of university campuses in reconnecting humans to the biosphere. *Sustainability, 9*(12), 2349.

Cotton, D. R. E., Miller, W., Winter, J., Bailey, I., & Sterling, S. (2015). Developing students' energy literacy in higher education. *International Journal of Sustainability in Higher Education, 16*(4), 456–473.

Cotton, D., Miller, W., Winter, J., Bailey, I., & Sterling, S. (2016). Knowledge, agency and collective action as barriers to energy -saving behaviour. *Local Environment, 21*(7), 883–897.

DeWaters, J., Qaqish, M., Graham, M., & Powers, S. (2013). Designing an energy literacy questionnaire for middle and high school youth. *The Journal of Environmental Education', 44*(1), 56–78.

Evans, J., & Karvonen, A. (2014). 'Give me a laboratory and I will lower your carbon footprint' Urban laboratories and the governance of low-carbon futures. *International Journal of Urban and Regional Research, 38*(2), 413–430.

Evans, J., Jones, R., Karvonen, A., Millard, L., & Wendler, J. (2015). Living labs and co-production: University campuses as platforms for sustainability science. *Current Opinion in Environmental Sustainability, 16,* 1–6.

Hogan, S., Pascale, A. Cetois, A., & Ashworth, P. (2019). *Building Australia's Energy Literacy*. Prepared for National Energy Resources Australia, The University of Queensland, n.d. https://www.nera.org.au/Featured-energy-literacy

Martins, A., Madaleno, M., & Ferreira Dias, M. (2020). Energy literacy: What is out there to know? *Energy Reports, 6*(Suppl. 1), 454–459.

McCrory, G., Schäpke, N., Holmén, J., & Holmberg, J. (2020). Sustainability-orientated labs in real-world contexts: An exploratory review. *Journal of Cleaner Production, 277*. https://doi.org/10.1016/j.jclepro.2020.123202

McKenna, E., Richardson, I., & Thomson, M. (2012). Smart meter data: Balancing consumer privacy concerns with legitimate applications. *Energy Policy, 41,* 807–814.

Senge, P. M. (2006). *The fifth discipline: The art and practice of the learning organization* (2nd ed.). Random House.

Sovacool, B. K., & Dworkin, M. H. (2015). Energy justice: Conceptual insights and practical applications. *Applied Energy, 142,* 435–444.

Wenger, E. (1999). *Communities of practice: Learning, meaning, and identity*. Cambridge University Press.

Open Access This chapter is licensed under the terms of the Creative Commons Attribution 4.0 International License (http://creativecommons.org/licenses/by/4.0/), which permits use, sharing, adaptation, distribution and reproduction in any medium or format, as long as you give appropriate credit to the original author(s) and the source, provide a link to the Creative Commons license and indicate if changes were made.

The images or other third party material in this chapter are included in the chapter's Creative Commons license, unless indicated otherwise in a credit line to the material. If material is not included in the chapter's Creative Commons license and your intended use is not permitted by statutory regulation or exceeds the permitted use, you will need to obtain permission directly from the copyright holder.

Index

© The Editor(s) (if applicable) and The Author(s) 2022
C. Howarth et al. (eds.), *Addressing the Climate Crisis*,
https://doi.org/10.1007/978-3-030-79739-3